The Book of the
BLACK FIVE
LM CLASS 5 4-6-0s
Part 1, 45000-45074

By
Ian Sixsmith

45044 runs light from Mold Junction towards Chester. The riveted tender is one of those changes which escaped the record and history cards; the last recorded tender was a welded pattern fitted in 1953. Photograph N. Kneale, www.transporttreasury.co.uk

Irwell Press Ltd.

ISBN 978-1-906919-40-5

First published in the United Kingdom in 2011
by Irwell Press Limited, 59A, High Street, Clophill,
Bedfordshire MK45 4BE
Printed by Short Run Press, Exeter

Contents

INTRODUCTION AND ACKNOWLEDGEMENTS   **Page 5**

1. **GO ANYWHERE AND DO ANYTHING – THE IDEAL LOCOMOTIVE**   **Page7**
   Stanier's Inheritance
   Early Deliberations
   The Scottish Dimension
   Authorisation
   Description

2. **HERE, THERE AND EVERYWHERE**   **Page 17**
   The end of steam and preservation

3. **THE DEVIL IN THE DETAIL**   **Page 23**
   Boilers
   Chimney and Top Feed Pipes
   Steam Pipe Casings
   Smoke Box Door
   Frames
   Wheels and Axles
   Cylinders and Inspection Covers
   Combination Levers
   Running Plate and Cab
   Carriage Warming Pipes
   Tablet Exchange Apparatus
   Snowploughs
   BR Days
   Tenders
   Liveries

4. **ON THE RECORD**   **Page 33**
   Health Warning
   Sheds...
   Repairs and Maintenance
   Works...
   Mileages
   'Improvements, Etc'
   Crosshead Driven Vacuum Pumps
   Steam Sanding
   Speed Indicators
   Modification and Modernisation
   AWS

Allan C. Baker writes: A Black 5 (we always referred to them as *Black Uns*) in the process of having its boiler washed out at Stoke in the 1960s. Notice the long extension on the water pipe (we called them *bags* and the metal extension a *lance*) which will pass through the wash-out plug holes at the bottom of the smokebox tubeplate. The wash out staff (we called them *boiler washers*) by dexterous use of this in the various holes, ensured that any scale and other detritus, accumulating at the bottom of the boiler barrel, was moved by the high pressure jet to the water spaces round the firebox. It could then be raked out through the wash-out hand-holes round the base of the outer firebox. We also have a clear view of the smokebox including the blast pipe at the bottom, petticoat pipe at the top, superheater elements and, to the extreme right, the vacuum brake ejector exhaust pipe making its way to a ring of jet holes between the petticoat pipe and the chimney. Notice too, the protection plate behind the screw coupling to protect the AWS receiver, which can also be seen with its connecting cable. The frequency of boiler washouts and the amount of scale removed depended on the hardness of the water which of course, varies from place to place. Some sheds therefore washed out boilers on a weekly basis and others fortnightly – engines working consistently in some parts of Scotland would last even longer. At Crewe it was every two weeks; thus we had WOX and BFX examinations, indicating an X day exam associated with a washout or a boiler-full. At some sheds with goods engines that had their X day examinations fortnightly, if the wash-out period was weekly, they would have to be stopped specially for a wash-out between X days. Photograph J. Vaughan, www.transporttreasury.co.uk

Introduction and Acknowledgements

45020, the very first Black 5, ready for departure from Watford in the early 1950s. At this time it was shedded at Rugby which it left in September 1953. www.rail-online.co.uk

When asked by my esteemed publishers for help in putting together a tome on the LMS Black 5s in the *Book Of* series the first question to be answered was, how many volumes? Surely even the mighty Irwell machine would struggle to cope with a thousand pager! After much discussion over sausage sandwiches and pints we came up with the answer – five. So, this Part 1 covers the 1934 batch from Vulcan Foundry and the 1935 engines from Crewe and Part 2 the similar 1935 Vulcan Foundry and Armstrong Whitworth locomotives. Part 3 will describe the 'Mark 2' 1936 Armstrong Whitworth locomotives and will sweep up the remaining pre-war engines. Part 4 will deal with the war-time and immediate post-war LMS batches leaving part 5 with the Caprottis and the final LMS and BR-built examples.

As we will discover, the Black 5s were not all the same – far from it – and I

trust the reader will follow through the story in the approximate chronological sequence which seemed at the time to make sense. The books therefore are arranged by the order in which the locomotives were introduced, with an added twist that, particularly in matters such as boilers and tenders, there is a certain amount of back and forward cross-referencing. In the belief that if you buy one book you will surely need the others to complete the set, some details are covered in more depth in the earlier books and only summarised in the later parts.

The core of the material by weight is from the Engine History Cards and Engine Record Cards aided and abetted by information begged and borrowed from a number of sources, and backed up by a very large pile of photographs. I would record my thanks especially to Paul Chancellor, Peter Groom, Michael Mensing, Norman Preedy, Rail-Online

and The Transport Treasury for allowing me to use their pictures. I owe special thanks too, to Allan C. Baker and Eric Youldon. I have also consulted magazines including *The Railway Gazette*, *SLS Journal*, *The Railway Observer* and the *LMS Journal*.

Finally, and in this case it is definitely last but not least, I could not have produced this book without the help of John Jennison of Brassmasters fame (Brassmasters – purveyors of exquisite etched brass kits of LMS prototypes including the Black 5s; PO Box 1137, Sutton Coldfield, West Midlands, B76 1FU; www.brassmasters.co.uk). He it was who analysed the hundreds of History and Record Cards and made available his extensive photograph collection, allowing me to fill in many gaps in the story.

An ex-LNWR 'Prince of Wales' 4-6-0 no. 5722 in the 1930s. There were several attempts to produce a modernised version of this class over the ten years before the first Black 5 was introduced. The LMS Board authorised the purchase of the Black 5s to replace these engines and the other ex-LNWR 4-6-0s. A total of 105 Prince of Wales were withdrawn in 1934-35 and another 105 in 1936-37 as more Black 5s entered service.

1. GO ANYWHERE AND DO ANYTHING
THE IDEAL LOCOMOTIVE

In 1939 the *Railway Gazette* under the heading L.M.S.R. GENERAL UTILITY LOCOMOTIVES observed: *The ideal locomotive, dreamed of and longed for by the operating departments of large railway systems everywhere, is one that could 'go anywhere and do anything.'* This, indeed, was the definition put forward, perhaps only half seriously, by the late Mr. J. H. Follows, then Chief General Superintendent of the L.M.S.R., when we discussed the subject with him in his office at Derby shortly after his appointment to that position in 1923. As we stated in the editorial columns of our issue of February 19, 1932, when recording his retirement, he pleaded for something in the nature of a 'universal' type of engine to dispose of the majority of the operating problems with which he had to deal. As we then pointed out, unfortunately for the full realisation of this idea, conditions are such as to make it impossible for the locomotive departments to provide such an engine owing to the numerous and conflicting operating circumstances associated with questions of gauge, track and structures, axle loadings, clearances of one kind or another, and the proper observance of economy in working. It is, however, possible nowadays, by the

exercise of designing skill and the most careful selection of dimensions, aided by the advances made in the metallurgical field, to produce a mixed traffic engine which, although not capable of giving entirely ubiquitous service, can and does go a long way in that direction. The factor that is new in the most recent designs is the increased range due to the ability of the modern engine to run at high speeds, a result of the improvements which have been made in valve-gear and steam port design. Hence what might formerly have been regarded as somewhat Utopian need no longer be dismissed as belonging to that category.

Locomotives are to be found on railways at home and abroad which cannot be said to rank precisely as belonging to any one classification; they are neither express passenger, freight, nor ' coal ' engines in the accepted meanings of the terms, but nevertheless by virtue of their design and proportions they can undertake, with reasonable efficiency the work of any of these. There are, of course, limiting circumstances which if used as a basis of comparison would sometimes show the engine at a disadvantage, and the term ' mixed traffic ' must not in this connection be confused with ' jack-of-all-trades ' or to be more exact, of all traffics.

A successful locomotive of this description provides a most valuable asset to a railway company, and when built in large numbers a considerable saving in first and maintenance costs results from the repetition methods applying to both production and repair. Among the factors to be considered in designing such an engine are (i) reasonable simplicity of design, (ii) compromise in the selection of dimensions, (iii) questions of varying loading gauges, track standards, axle loadings, &c., and (iv) extended engine workings; and although at first sight some of the factors may appear in a measure irreconcilable, practice shows that in modern circumstances this is no longer prominently so.

Investigation of this subject in the abstract is made easier when associated with a study of concrete cases of locomotives planned to meet mixed traffic requirements. An excellent example of modern practice is found in the Class '5' 4-6-0 engines introduced by Mr. Stanier on the L.M.S.R. in 1935. Since that year nearly 500 have been built, representing a very considerable capital cost; the engines have, however, proved so successful in working 'under a wide range of service conditions on virtually all parts of the company's system, so

The official portrait of the first Black 5 clearly shows off the features peculiar to the first batch built. From front to back these are: no steam heat pipe or front platform cover, tall chimney, the oval Vulcan Foundry worksplate above the steam pipe, the prominent top feed pipes, the absence of a rain gutter on cab roof and the plain tender axlebox covers. Other items which either changed or disappeared during the early years as the design evolved were the hollow bogie axles, cranked combination lever, crosshead driven vacuum pump and strengthening webs on the coupled wheels. This picture also shows the 40in spacing of tender letters and lining on the cab from the top front corner a short way along the eaves, down the front edge of the front window, below the windows, up the rear edge of the rear window and along the eaves to the top rear corner.

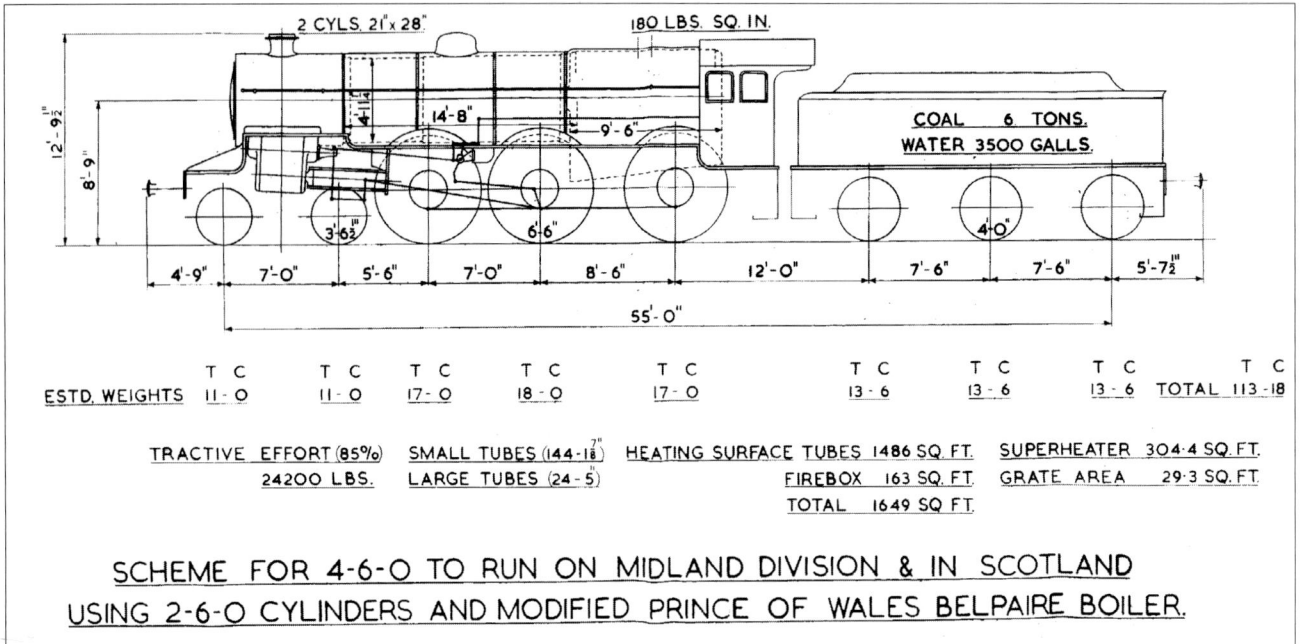

	T C		T C	T C	T C		T C		T C		T C		T C			T C
ESTD. WEIGHTS	11 - 0		11 - 0	17 - 0	18 - 0		17 - 0		13 - 6		13 - 6		13 - 6		TOTAL	113 - 18

TRACTIVE EFFORT (85%)   SMALL TUBES (144 - 1⅞")   HEATING SURFACE TUBES 1486 SQ. FT.   SUPERHEATER 304·4 SQ. FT.
24200 LBS.   LARGE TUBES (24 - 5")   FIREBOX 163 SQ. FT.   GRATE AREA 29·3 SQ. FT.
TOTAL 1649 SQ. FT.

SCHEME FOR 4-6-0 TO RUN ON MIDLAND DIVISION & IN SCOTLAND USING 2-6-0 CYLINDERS AND MODIFIED PRINCE OF WALES BELPAIRE BOILER.

The 1924 proposal for an outside cylindered 4-6-0 using the cylinders from the forthcoming Crab 2-6-0, which was being finalised, and a modified 'Prince of Wales' Belpaire boiler. It was intended to meet the tight weight limits on many lines in northern Scotland and the north west of the Midland Division but did not come to fruition, leaving the Northern Division to wait for a decade before it had any modern motive power.

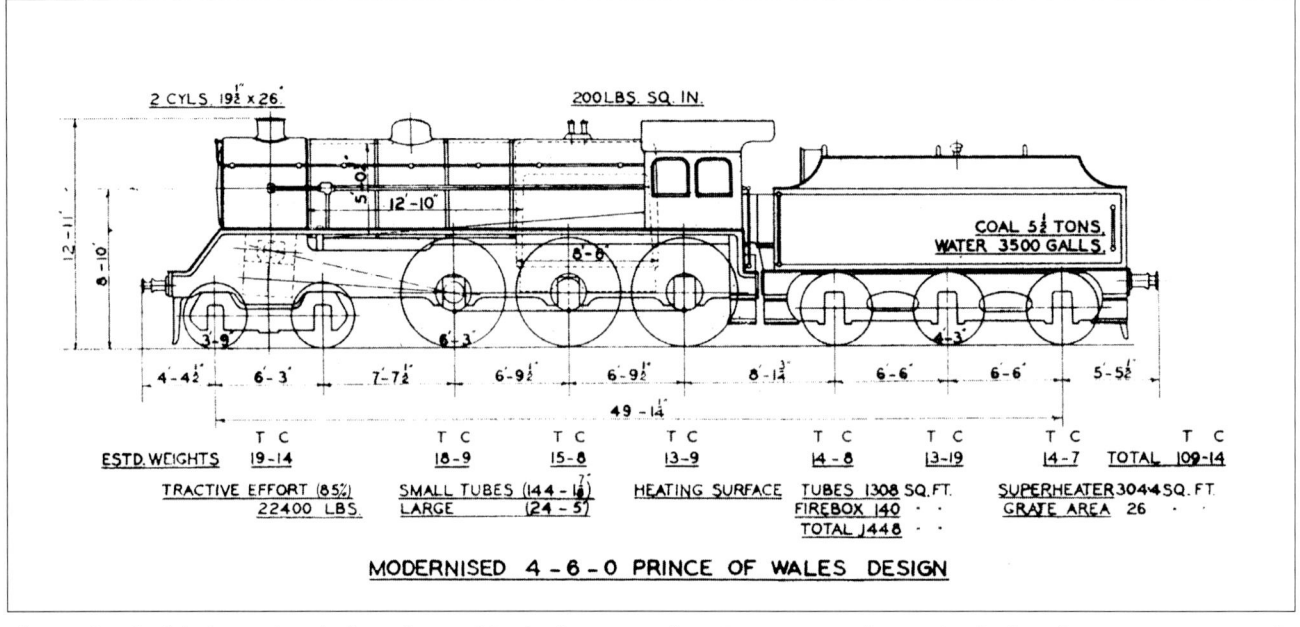

	T C	T C	T C	T C		T C	T C		T C			T C
ESTD. WEIGHTS	19 - 14	18 - 9	15 - 8	13 - 9		14 - 8	13 - 19		14 - 7		TOTAL	109 - 14

TRACTIVE EFFORT (85%)   SMALL TUBES (144 - 1⅞")   HEATING SURFACE TUBES 1308 SQ. FT.   SUPERHEATER 304·4 SQ. FT.
22400 LBS.   LARGE (24 - 5")   FIREBOX 140 · ·   GRATE AREA 26 · ·
TOTAL 1448 · ·

MODERNISED 4 - 6 - 0 PRINCE OF WALES DESIGN

The Modernised 'Prince of Wales' put forward in the late-1920s by HPM Beames, the Mechanical Engineer at Crewe, was in many ways ahead of its time with its Caprotti valve gear, albeit worked from inside the frames, a high running plate and cut-off cab. The scheme was dusted off again in 1931, with Walschaerts outside gear, and formed the basis for the design included in the 1933 Locomotive Programme which was subsequently cancelled and replaced by the Black 5.

economical in fuel and maintenance, and so generally satisfactory, that the investment must be regarded as excellent. From the strictly locomotive point of view the Class '5' series are admirable alike in respect of their undoubted efficiency, compactness, symmetry of appearance, and ubiquity of service.

The article continued to enthuse: The L.M.S.R. Class '5' 4-6-0 mixed traffic engines, on account of their operating economy and wide range of usefulness, conform very closely to this designation [General Utility Locomotives]. The fact is well established that large railway systems, operating extensive and widely diversified classes of traffic must, of necessity, have at their disposal a sufficiency of locomotives conforming to the 'general utility,' or as it is usually termed mixed traffic category, in order that undue multiplication of engine types may be avoided and the economies attending standardised workings realised.

The task of designing a locomotive of this character entails the most careful selection of dimensions, and due regard must be paid to factors which do not arise in such an acute form when planning engines for a specified and more clearly defined form of duty consisting almost entirely of a unified type of service. The wide difference between an express passenger and a heavy 'coal' engine, for instance, has in large measure to be bridged by the mixed traffic class, and the latter, to achieve its purpose, must effectively further the policy of extended engine workings by making it possible to work successively, with the same engine, trains of the fast passenger, fitted freight and even those of the loose coupled type. Mixed traffic engines have, of course, been designed since the earliest times, but the factor that is new in the most recent designs is the increased range due to the ability of the modern engine to run at high piston speeds. This in turn is attributable to the improvements which have been made in valve gear and steam port design.

The subject is of considerable interest and importance to the operating departments of

STANDARD DESIGNS		1923	1924	1925	1926	1927	1928	1929	1930	1931	1932	Totals
Class 6	4-6-0 Pass.Tender (Royal Scot)					50			20			70
Class 5X	4-6-0 Pass.Tender (Patriot)								(2*)		15	15
Class 4	4-4-0 Pass.Tender (Compound)		40	95	5	50					5	195
Class 2	4-4-0 Pass.Tender						50	19	4	30	35	138
Class 4	2-6-4 Pass.Tank					4	21	50			10	85
Class 3	2-6-2 Pass.Tank								21	39	10	70
Class 2	0-4-4 Pass.Tank										9	9
Class 7	0-8-0 Freight Tender							100	3	32	40	175
Class 5	2-6-0 Freight Tender				13	87	8	22	95	10	10	245
Class 4	0-6-0 Freight Tender		11	161	132	137	89					530
	2-6-6-2 Freight Tank (Garratt)					3			30			33
Class 3	0-6-0 Freight Tank		42	8	128	36	157	36		15		422
Class 2	0-6-0 Freight Tank (Dock)						7	3				10
Class 0	0-4-0 Freight Tank (Saddle)										5	5
TOTAL STANDARD DESIGNS			93	264	278	367	332	230	173	126	139	2,002
EXISTING NON-STANDARD DESIGNS PERPETUATED												
Class 5	4-6-0 Pass.Tender (4-cyl L&YR Class 8)	21	16	4								41
Class 4	4-6-0 Pass.Tender (LNWR Prince of Wales)		1									1
Class 4	4-6-0 Pass.Tender (Caledonian 14630 Class)			2	18							20
Class 3	4-4-2 Pass.Tank (LTSR)	10		5		10			10			35
Class 2	0-4-4 Pass.Tank (Caledonian)			10								10
Class 7	2-8-0 Freight Tender (SDJR)			5								5
Class 3	0-6-2 Freight Tank (North Stafford)	4										4
TOTAL NON-STANDARD DESIGNS PERPETUATED		35	17	26	18	10	0	0	10	0	0	116
NON-STANDARD NEW DESIGNS												
Class 5	4-6-4 Pass.Tank (4-cyl Baltic)		10									10
Class 7	0-8-4 Freight Tank (Shunter)	29	1									30
TOTAL NON-STANDARD NEW DESIGNS		29	11	0	0	0	0	0	0	0	0	40
"SENTINEL" LOCOMOTIVES								2	4		1	7
GRAND TOTAL		64	121	290	296	377	332	232	187	126	140	2,165
* These two engines were counted as rebuilds and are not represented, therefore, as additions to stock												

New locomotives added to LMS stock 1923-1932

railways and of no less moment to the locomotive department on which latter rests the responsibility for the provision and maintenance of adequate and suitable engine power.

Stanier's Inheritance
By the time William Stanier joined the LMS in January 1932 the company had already built a large number of new locomotives in an attempt to eliminate 'existing engines of inferior performance and efficiency' and had reduced the number of classes by almost half, from 393 in 1923 to 230. Naturally as the largest constituent companies, the Midland Railway and LNWR engines bore the brunt of this reduction with around 40% of the inherited stock from each taken out during this period. The table below shows the new locomotives added to stock during the first ten years following the grouping; no Stanier designs emerged until 1933 so it forms a convenient cut-off point. The 2,165 locomotives were primarily based on Midland Railway designs and, with the exception of the 245 Hughes-Fowler 'Crab' 2-6-0s, none of these were mixed traffic locomotives as defined by the *Gazette*.

Early Deliberations
The only significant number of 4-6-0s in the LMS stock were the LNWR 'Experiment', 'Prince of Wales' and '19 inch Goods', none of which scored well against Follows' definition of a mixed traffic locomotive. The Midland's contribution was the 4F 0-6-0 and even though some were used on passenger work on secondary lines they were and remained primarily freight engines. As early as 1924 a design had been arrived at to marry a modified 'Prince of Wales' Belpaire boiler with the cylinders from the forthcoming Crab 2-6-0 to produce an outside cylindered 4-6-0 to run on the Midland Division and in Scotland, but this fell by the wayside and only the 2-6-0s were built.

At Crewe the ex-LNWR men thought that the 'Prince of Wales' could be modernised and its many problems such as frame fractures eliminated. H.P.M. Beames, the Mechanical Engineer at Crewe, put forward a proposal for a modernised 'Prince of Wales' which had inside cylinders with outside valves driven by Caprotti gear and a very futuristic high running plate and cut-off cab, all of which would eventually re-appear in the final pair of British Railways-built Caprotti Black 5s. Nothing

came of this until after Beames had moved to Derby as Deputy CME and the result was a 'superheated converted Prince of Wales' with outside Walschaerts valve gear. Ten of these were included in the 1933 Locomotive Renewal Programme with the justification: 'Having regard to the heavy Passenger stopping trains now worked by Prince of Wales type engines, these engines are heavy on coal and are falling due for re-boilering. It is suggested that they be replaced by a similar engine of modern design'. However the LMS had to wait another year for its 'go-anywhere and do anything' 4-6-0 because the Programme was amended in May 1933 before construction began. It would not be possible to complete the designs 'to ensure their being built this year'. They were replaced by additional '3-cylinder Claughton 4-6-0s' [Patriots], and deferred to the 1934 Programme.

The Scottish Dimension
In parallel with Beames' proposals were growing demands from the Northern Division for a powerful modern locomotive which could operate within the restricted Scottish axle loads. The LMS had inherited several classes of

9

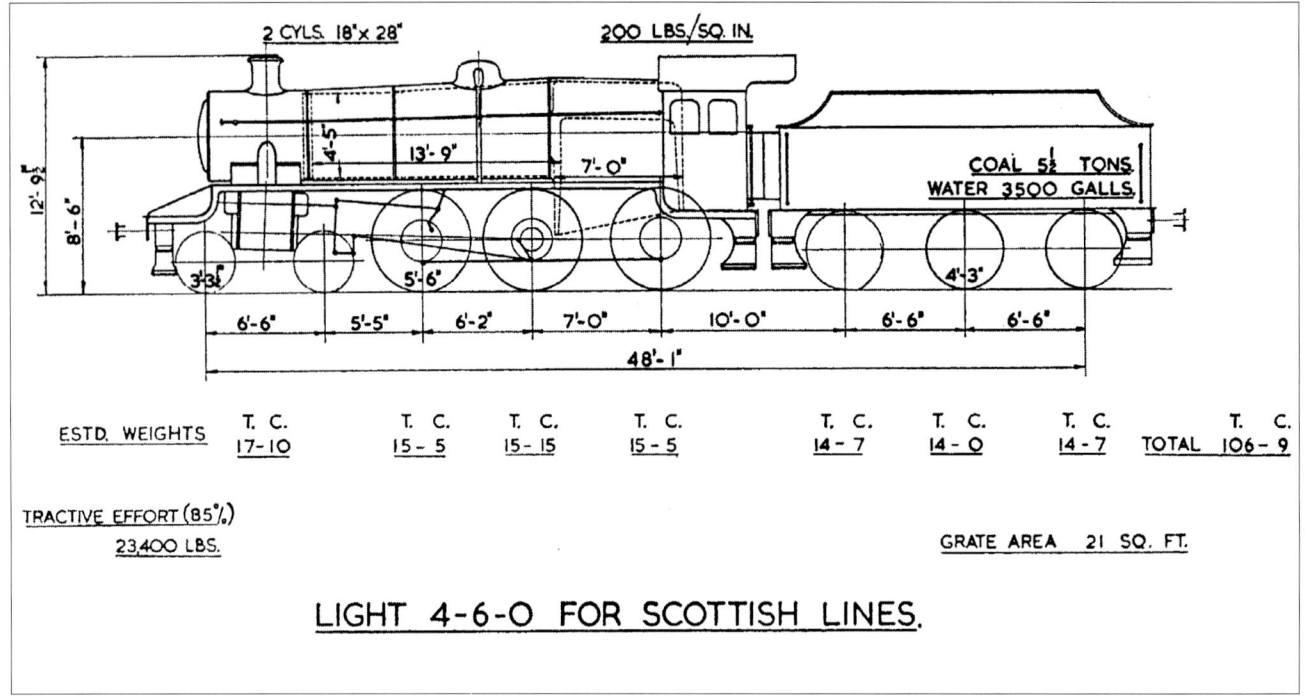

LIGHT 4-6-0 FOR SCOTTISH LINES.

The lightweight 4-6-0 intended to produce a modern powerful locomotive able to meet the weight restrictions of the LMS Northern Division. Ten were approved in the 1934 Locomotive Programme but like the Modernised 'Prince of Wales' were cancelled and replaced by the Black 5s.

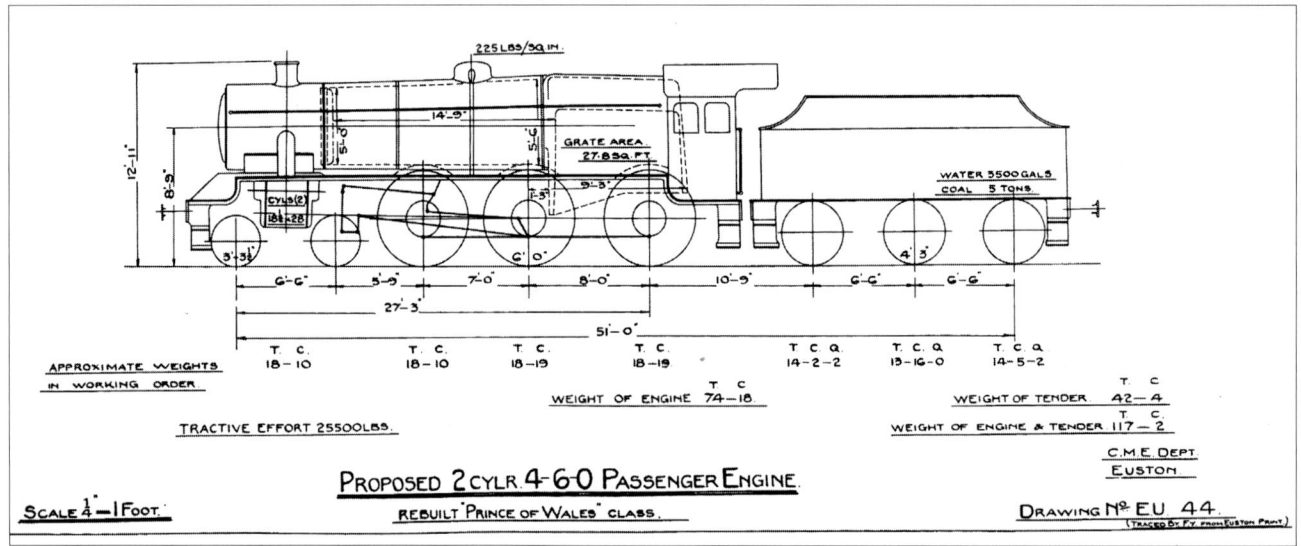

Diagram EU 44, described as a Rebuilt 'Prince of Wales' class, was one of a number of outline designs issued on 8 July 1932 over Stanier's signature. This replaced two versions of a 3-cylinder taper boiler 4-6-0 with 6ft 0in driving wheels on a 7ft 0in + 8ft 0in wheelbase to replace the 'Prince of Wales' Class. Both were discarded due to weight considerations and this version with two 18½ in x 28in cylinders and vertical throatplate boiler, was the basis for the final Class 5 proposal.

4-6-0 designed by each constituent company to meet its local conditions, but by the 1930s these were ageing and struggling to cope with traffic demands. The 1924 design had come to nought and the only new engines subsequently introduced north of the border were Class 2P and Compound 4-4-0s and the 4F 0-6-0s, plus ten Crab 2-6-0s which went to the Highland Section during 1928/29, though these were restricted from many of the far-north lines.

Serious consideration was given to re-boilering existing Scottish classes to prolong their life and this was summarised in a statement dated September 1931, *LMS - Particulars &C. Classification of Engines on Northern Division* which listed all of the Division's

locomotives with their construction date, the date its boiler was new, whether it was fitted with a standard Northern Division boiler, if it could be fitted with such a boiler, the standard LMS class which was a suitable replacement, and a recommendation for its future. The Caledonian Railway Class '60' 4-6-0s came under the category 'useful type and fitted with ND standard boiler. Should continue in service'. Several classes were designated 'Limited axleload, and must be retained in traffic or replaced by similar light engine until these limits on axleloading are removed. May be reboilered'. On the Highland Section these were the HR 'Clan', 'Castle', 'River', and 'Jones Goods' 4-6-0s, on the Kyle line the HR

Cummings 'Superheated Goods' 4-6-0s and for the Callander & Oban Section, the CR '55' 4-6-0s and Pickersgill '191' 4-6-0s.

No decision had been made before Stanier arrived from the GWR and discussions continued into 1933 with the dilemma. Should some existing types be reboilered or should a new lightweight 4-6-0 be introduced, which might not be powerful enough to avoid double-heading on some sections of the Highland main line? As time passed, the problem was getting worse because the boilers on the pre-group 4-6-0s were ageing and more were coming up for renewal. The proposal which finally won the day was a scaled-down version of the 4-6-0 with an estimated maximum

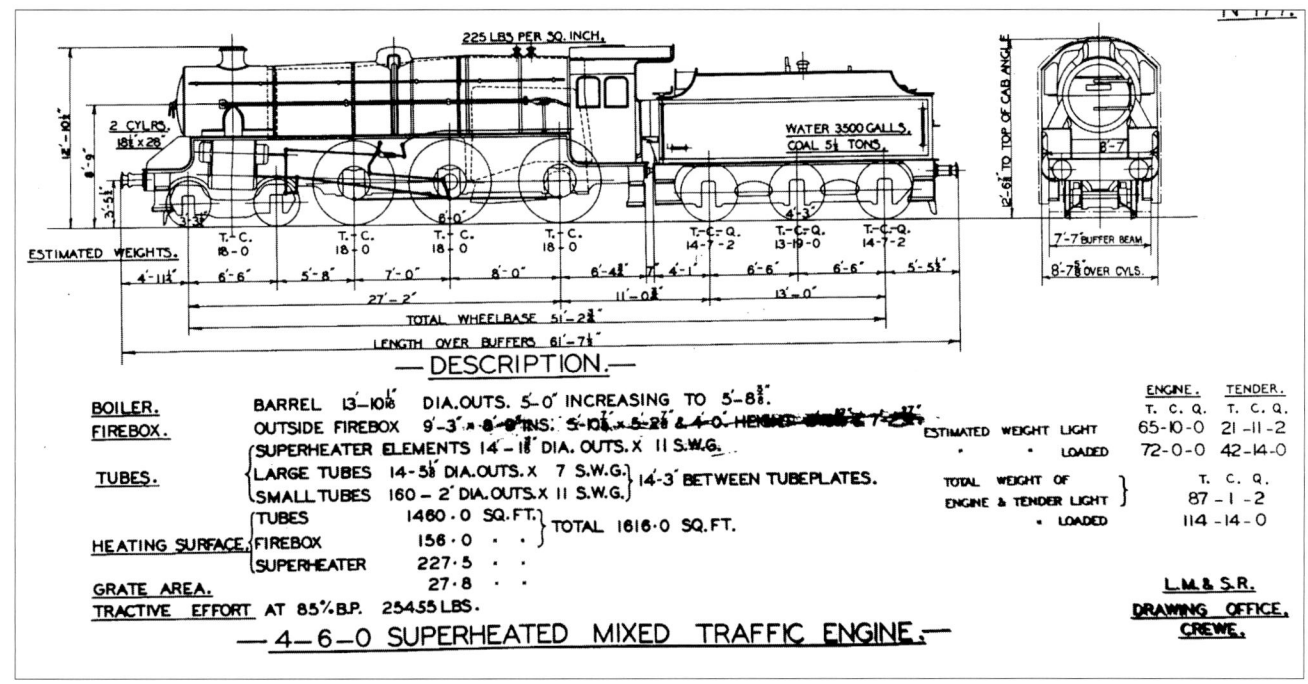

225 LBS PER SQ. INCH.

2 CYLRS. 18½"×28"

WATER 3500 GALLS. COAL 5½ TONS.

12'-6½" TO TOP OF CAB ANGLE

8'-7"

7'-7" BUFFER BEAM
8'-7¾" OVER CYLS.

ESTIMATED WEIGHTS.

| T.C. | T.C. | T.C. | T.C. | T-C-Q. | T-C-Q. | T-C-Q. |
| 18-0 | 18-0 | 18-0 | 18-0 | 14-7-2 | 13-19-0 | 14-7-2 |

4'-11¼"   6'-6"   5'-8"   7'-0"   8'-0"   6'-4½"   4'-1"   6'-6"   6'-6"   5'-5½"

27'-2"   11'-0½"   13'-0"

TOTAL WHEELBASE 51'-2¾"

LENGTH OVER BUFFERS 61'-7½"

— DESCRIPTION.—

BOILER. BARREL 13'-10⅛" DIA.OUTS. 5'-0" INCREASING TO 5-8⅜".
FIREBOX. OUTSIDE FIREBOX 9'-3" × 8'-9" INS. 5'-10⅜" × 5'-2⅝" & 4'-0" HIGH 7'-2".
TUBES. { SUPERHEATER ELEMENTS 14 – 1⅜" DIA. OUTS. × 11 S.W.G.
        { LARGE TUBES 14 – 5⅛" DIA.OUTS. × 7 S.W.G. } 14'-3" BETWEEN TUBEPLATES.
        { SMALL TUBES 160 – 2" DIA.OUTS. × 11 S.W.G. }
HEATING SURFACE. { TUBES   1460·0 SQ.FT. }
                 { FIREBOX   156·0 · · } TOTAL 1616·0 SQ.FT.
                 { SUPERHEATER   227·5 · · }
GRATE AREA.   27·8 · ·
TRACTIVE EFFORT AT 85% B.P. 25455 LBS.

	ENGINE.	TENDER.
	T. C. Q.	T. C. Q.
ESTIMATED WEIGHT LIGHT	65-10-0	21-11-2
" " LOADED	72-0-0	42-14-0
TOTAL WEIGHT OF ENGINE & TENDER LIGHT }	T. C. Q.	87-1-2
" " LOADED		114-14-0

L.M.& S.R.
DRAWING OFFICE.
CREWE.

—4—6—0 SUPERHEATED MIXED TRAFFIC ENGINE.—

ED 177 was the final design proposal with Fowler 3500 gallon tender. Other than substituting the new Stanier 4000 gallon tender, this was the version actually built.

axle load of 15tons 15cwt which would allow it to be used on the Callander and Oban line, although it would probably not have been powerful enough for the heaviest duties on the Highland.

Authorisation
During the first few months of 1932 Stanier's team had sketched out a range of new locomotives from which specific ones were selected for detail design work at the drawing offices. This included two versions of 3-cylinder taper boiler 4-6-0 to replace the 'Prince of Wales' Class, both having 15½in x 28in

cylinders and 225 psi boiler pressure producing 26,800 lb of tractive effort, one with a sloping throatplate and the other a vertical throatplate boiler. Both were too heavy for general use over the majority of the system, especially the Midland and Northern Divisions where restrictions were most severe, and another version with two 18in x 28in cylinders and vertical throatplate boiler was produced in July 1932 (diagram EU 44). This was the basis of the final Class 5 proposal put forward to the Board.

The LMS 1934 Locomotive Renewal Programme approved in June 1933

included ten of these 'Improved Prince of Wales' 4-6-0s; the frames of the existing locomotives gave a great deal of trouble and the cost of maintenance and coal was heavy. The new locomotives, it was declared, 'would be of 17% increased tractive power, whilst the cost of maintenance and coal would be about 11% less'. Scotland had also finally secured its new engines with ten Northern Division 4-6-0 passenger locomotives approved: *The present locomotives had reached the end of their theoretical life, and it would not be economical to reboiler them. They were*

With the paint on its buffers still unblemished 5020 poses at Crewe in August 1934. It has not yet been fitted with a worksplate and although officially allocated to Perth from 2/8/34 it remained in England and was used for tests during September. It went to Carlisle in October before returning south to Edge Hill the following November.

Four of the five 1935 Crewe-built batch, 5070-74, can be seen under construction in the Erecting Shop at Crewe Works in April 1935. The one at the front is 5070 identified by E393 -1 chalked on the smokebox rim.

saturated engines, heavy on coal, and not satisfactory in service, and the new locomotives to replace them would reduce maintenance and coal costs by about 6%. The capacity would remain about the same, as the weight was restricted by physical considerations.

However, once again, things were not to proceed according to plan; within a few weeks the LMS had determined upon a 'scrap and build' policy. The Minutes of the October 1933 Board Meeting record the following: *The Chairman referred to the memo dated October 1933 giving particulars of the estimated stock of locomotives at 31st December 1934 and recommending that 121 locomotives, the boilers of most of which would require to be renewed shortly or which otherwise would not in the ordinary course be broken up for some years, be replaced by one hundred engines, fifty of which would be of the 'Improved Claughton' [Jubilee] type and fifty the improved 'Prince of Wales' [Black 5] type. This would result in an 18% saving in coal consumption and 25% saving in cost of repair, per mile. It was agreed that the Locomotive Trade be asked to submit alternative tenders for each type in lots of twenty-five and fifty locomotives.*

In November 1933 quotes were invited for 4-6-0 2-cylinder and 3-cylinder locomotives and tenders, in lots of 25 or 50. Eight different firms tendered for the 2-cylinder type, with

prices ranging from £5,625 to £6,900 for lots of 25, and four of these also tendered for lots of 50 at prices between £5,485 and £6,795. (The same firms tendered for the 3-cylinder type at prices ranging from £5,810 to £7,525 for lots of 25, and from £5,710 to £7,400 for lots of 50). Fifty of each type were ordered, the North British Locomotive Company winning the 3-cylinder engines at £5,710 each. Although this firm's tender for the 2-cylinder lot was the lowest by £55 the contract was awarded to Vulcan Foundry at £5,540 each. The need for new locomotives quickly, and especially to solve the Northern Division problem, meant the additional cost of £2,750 was a price worth paying to accelerate deliveries. If NBL had won both orders the delivery would have extended over 54 weeks, but by dividing them delivery of 37 weeks from Vulcan and 49 weeks from NBL could be achieved. This was ratified at the next Board meeting in November. Tenders had been received from eight firms and 'it was resolved that the tender of the Vulcan Foundry Company Ltd of Newton-le-Willows for fifty 4-6-0 superheated 2 cylinder mixed traffic engines at a price of £5,540 each be accepted'.

Almost inevitably, the lightweight 4-6-0 for Scotland was finally abandoned as civil engineering improvements, particularly on the Oban line, would

allow the new design to be widely used on the Northern Division. Accordingly in April 1934 Stanier and the Chief Operating Manager recommended that *twenty 4-6-0 engines and tenders of a new type suitable for mixed traffic work both in England and Scotland be built in the Company's workshops in place of the ten improved 4-6-0 'Prince of Wales' type and the ten 4-6-0 engines and tenders for the Northern Division authorised... as part of the Locomotive Renewal Programme for 1934, and that in order to give a wider use of the engine, the tenders be of 4000 gallons capacity, similar to the 50 engines of this type ordered from the Vulcan Foundry.* These were to be built in LMS Works and emerged as 5000-5019, but the first was not delivered until the month following the final engine of the fifty from Vulcan (5020-5069).

The 1935 Renewal Programme adopted in June 1934 included another 55 mixed traffic 4-6-0s and tenders at a cost of £338,250 of which five were to be built at Crewe (5070-5074). The other fifty were ordered from Vulcan Foundry (5075-5124) together with another hundred from Armstrong Whitworth (5125-5224), but we are getting ahead of ourselves here and these engines will all be dealt with in Part 2.

Description

In July 1934 the CME's Department

issued a press release entitled DESCRIPTION OF 2-CYLINDER

Lot	Nos.	Works nos.	Built at	Dates
114	5000-5019	216 –235	Crewe Works	Feb 1935-May 1935
119	5020-5069	4565-4614	Vulcan Foundry	Aug 1934-Jan 1935
122	5070-5074	236 – 240	Crewe Works	May 1935-Jun 1935

MIXED TRAFFIC TENDER ENGINES.

Included in the 1934 building programme are 70, 2-cylinder 4-6-0 superheated mixed traffic engines, 50 of which will be built by the Vulcan Foundry and 20 at the Railway Company's Works, Crewe.

The engines will be numbered as follows:-
20 engines built at Crewe, 5000 to 5019
50 engines built by the Vulcan Foundry Ltd., 5020 to 5069

These engines have been designed to meet the requirements for a general utility engine. The general appearance and principal dimensions are shown on the attached photograph and diagram.

BOILER

A Belpaire type of boiler with a taper barrel and a working pressure of 225 lb/sq.inch has been provided, and with a view to reducing the weight, 2% nickel steel plate has been used.

SUPERHEATER

A superheater is fitted, and the main regulator has been incorporated in the superheater header in the smokebox.

CONTROLS FOR STEAM SUPPLY

A steam manifold (with a main shut-off valve) is provided at the top of the firebox doorplate in the cab, on which are attached the necessary valves for the following:
Ejector and steam brake
Injectors
Carriage warming
Whistle
Pressure gauge
Sight feed lubricator to regulator

BOILER FEED

The feed water is supplied through top feed valves provided on the second boiler barrel ring with water distributing trays.

SAFETY VALVES

Two pop type safety valves (2½" diameter) are fitted at the crown of the firebox.

FIREDOOR

This is a standard type of sliding firedoor with a screen to prevent glare from the fire.

INJECTORS

An exhaust steam injector (10 m/m cones) is fitted on the fireman's side (right hand side) and on the other side a live steam injector (10 m/m cones) is fitted.

GENERAL FITTINGS

Other boiler mountings, such as the water gauge frames and protectors, etc., are of the Railway Company's standard type.

CYLINDERS

The two cylinders, which are carried outside the frames, are 18" diameter x 28" stroke.

MOTION

Walschaert type motion is fitted, the valve travel being 6½".

LUBRICATION FOR CYLINDERS

The piston valves, cylinders, piston rod packing and valve spindle bushes are provided mechanical lubrication, while the oil to each piston head is steam atomised. The mechanical lubricators are the Railway Company's standard type.

COUPLING AND CONNECTING RODS AND MOTION

These, in accordance with the latest practice, are of high tensile manganese molybdenum steel, the connecting rods being of a fluted section, but the coupling rods are of rectangular section.

WHEEL CENTRES

These are steel castings with the wheel rim of triangular section, and the tyre fixing is of the 'Gibson' retaining ring type.

The balance weights for the coupled wheels are built up by steel plates on both sides of the spokes, and riveted, the requisite weight being provided by filling in between the plates with lead.

AXLEBOXES FOR COUPLED WHEELS

These are steel castings with pressed-in

The class underwent several bouts of testing in the first twelve months as the deficiencies of the early low superheat boilers became apparent. 5051 was used for tests on the Western Division main line in March 1935. On the 19th, fitted with this impressive indicator shelter, it worked the 12.44pm from Crewe to Euston, returning with the 10.30am from Euston the next day. On the 2nd it worked the 12.44pm from Crewe again but this time returned the following day on the 5.30pm from Euston.

13

This is one of those official photographs which was not what it seemed. 5000 did not have a standard 4000 gallon tender and its firebox did not have washout inspection doors with small domed covers, there are no strengthening webs on the coupled wheels and it has plain cylinder wrappers without an access cover. It all points to a renumbering for the photographer of one of the 5070-74 batch. The works grey shows the Crewe pattern of cab lining carried straight up to the roof, and the 60in spacing of LMS on the tender.

brasses having suitable white metal crowns. Oil grooves are provided on both sides of each crown to ensure a thorough distribution of oil to the journal. The coupled axleboxes are arranged so that the oil pads can be examined by sliding out the underkeep while the axlebox is in position.

Each axlebox is fitted with a dust shield carried on the inside face of the box.

AXLEBOX LUBRICATION
A mechanical lubricator supplies the coupled axleboxes, each of which has an independent oil feed to the crown of the box, with the standard back pressure valve and flexible oil pipe connection.

SPRINGS AND GEAR
All the laminated bearing springs for the engine and tender are made of silico manganese steel, the plates being of ribbed section, and with the cotter type fixing in the buckle.

The spring links are of the screwed adjustable type.

FOUR WHEELED BOGIE
This is the standard type of four-wheeled bogie, the weight being taken through side bolsters, and bogie side check spring gear is provided to ensure smooth riding.

CAB
The width over the cab is 8'6".

The drive is on the left hand side, and all controls are arranged for convenient handling.

A tip-up seat is fitted on each side of the cab.

Two sliding windows are fitted on each side of the cab, and hinged windows are suitably arranged on the cab front plate.

On each side of the cab a small hinged window is fitted, which acts as a draught preventer for the enginemen when looking out.

BRAKE
Steam brake is provided on each side of the coupled wheels, and operated by the Driver's vacuum brake valve.

A vacuum pump is provided on the left hand side of the engine, this being suitably driven from the crosshead.

MECHANICAL SANDING
This apparatus is of the mechanical trickle type, the sand being delivered as follows:-
Front of leading coupled wheels.
Front and back of middle couple wheels.

In addition to this, water de-sanding apparatus is embodied which automatically comes into action, so that after the engine

has used the sand in the fore or reverse direction, as the case may be, the rails are cleaned with hot water to prevent interference with the track circuits.

CARRIAGE WARMING
The standard carriage warming apparatus is provided.

TENDERS
4,000 Gallon Tender
The tender is of the 6-wheeled type, the wheelbase being 15'0" and the coal capacity 9-tons.

The coal bunker provided has been carefully arranged so that as far as possible the coal will be self-trimming.

COAL BUNKERS
A bunker door is provided to give access to the coal space from the engine footplate, and ample toolbox space is provided at the front of the tenders.

FIRE IRONS
On the left-hand side of the tender, a suitable cavity is arranged in which are housed all the necessary firing irons.

WATER PICK-UP AND HAND BRAKE GEAR
Both the water pick-up and tender hand brake handles are arranged vertically. Bevel wheels

are provided for transferring the motion to their respective gears.

STEAM BRAKE
A steam brake is provided on each side of the six tender wheels, and is applied simultaneously with the steam brake on the engine.

INTERMEDIATE DRAWGEAR
As a further means of obtaining a smooth riding locomotive, the intermediate drawgear has been carefully designed with buffing spindles controlled by coiled springs.
The buffing spindles have each a specially designed head, which rides on an inclined plane provided on the hind engine buffer beam.

Wheel Base and overall length
Total wheel base, engine and tender = 53ft - 2¾in
Total length over buffers = 63ft - 7¾in

5020 as built showing the additional lining around the cab windows compared with the Crewe-built engines. Other differences include the close spaced tender lettering, external feed pipes to the top feed clacks and plain tender axlebox covers.

5013 at its home shed of Inverness on 8 September 1937. By this date the class was virtually omnipotent and had taken over the majority of services in the far north of Scotland. Attached to the rear left hand cabside, 5013 has automatic tablet exchange apparatus for working on the predominantly single lines. When not in use, the jaws were held upright against the side plate and as the tablet was to be exchanged they would be swung downward through 90 degrees to project the required distance from the engine to make the exchange. Photograph R.K. Blencowe.

5042 at Elstree in 1937 with the 4.25pm down Manchester, It was one of a number of Black 5s transferred to the Midland Division in February 1935, moving to Trafford Park. Those at Trafford Park displaced Compounds and along with engines from Kentish Town began working the Manchester (Central) to St.Pancras expresses. At that time they were the largest engines allowed over this line because of weight restrictions over bridges at Chapel-en-le-Frith.

2. HERE, THERE AND EVERYWHERE

With the motive power situation in Scotland it was not surprising that priority was given to the Northern Division when the first Black 5s were delivered from Vulcan Foundry in the summer of 1934. 5020 was kept back for testing while 5021-5029 went to Perth North. The remainder were shared out across the other Divisions and by January 1935 the Western Division had 23 engines, the Central and Northern Divisions ten each and the Midland Division seven. The latter must have been deemed in greater need of the new engines because the next month it received 17 from the Western and 3 from the Central Division. It quickly put them to use on both the West of England line to Bristol as well as the main line from St.Pancras.

Scotland then got its hands on nineteen of the twenty, 5000-5019, which emerged from Crewe Works between February and May. There was an indication of the harder duties performed north of the border when in the autumn the Northern Division exchanged its original 14-element superheater boiler engines, 5020-5029 and 5000-5006, for later ones with 21-element superheaters. (A similar move occurred in late-1936 when the Midland Division received thirty of the Armstrong Whitworth 5225-5451 series

to replace, in the main, the earlier Vulcan built engines which went to the Western Division).

The *Railway Observer* summed up the impact in Scotland: *The most noteworthy development on the H.R. this year is the introduction of Standard 2-cyl 4-6-0s of the '5000' series. There are now twenty of these locos on the section: thus outnumbering any other single class. [Even prior to 1923 there were only 19 Castles in service]. The new locos are to be seen on practically all of the express trains between Perth, Aviemore and Inverness by both the Carr Bridge and Forres routes, and are common between Inverness and Wick. Only the Skye line is free from these, and there the Jones Goods of 1894 share the work with the Cumming 4-6-0s of 1917-1919. Excellent work is also done by Highland engines between Inverness, Forres, Elgin and Keith, where the 'Wee Bennies' handle the through Aberdeen expresses to and from the 'Great North'. On the main line it is a different story. The 'Clans' have all disappeared. The 'Rivers' and the Horwich design of 2-6-0s are chiefly in demand when a Stanier 4-6-0 requires pilot assistance between Perth and Inverness, though this apparently is only necessary when the load exceeds about ten standard corridor coaches, despite the 1,484 feet of Drumochter.*

Elsewhere, the class settled down to perform the wide range of duties

expected of a modern mixed traffic type combining express passenger, fitted freight and other work with ease. Within two years of their introduction they had succeeded in reaching points to the north, south, east and west of the LMS empire. One of the last places they conquered was the Somerset & Dorset Joint where weight restrictions over several bridges between Mangotsfield and Bath prevented their use over the line until remedial work had been completed. Although work started in 1935 it was not until October 1937 that the Black 5s were cleared to work between Mangotsfield and Bath, subject to a speed restriction of 15 mph over two bridges. They soon appeared on excursions from the North and Midlands, but it was not until May 1938 that regular workings began and from this date these trains were worked by Bath and Saltley engines on alternate days. In preparation for this, six engines were transferred to Bath including 5023 from Upperby and 5029 from Shrewsbury.

With the outbreak of the second World War in September 1939 passenger services were cut back severely and therefore more express engines were available, which meant there was less need for mixed traffic engines to work passenger trains. This

Kentish Town's 5053 in 1936 at Leigh on Sea with ancient suburban stock working from Southend to St Pancras, as indicated by the destination board above the front buffer beam. 5053 is still in original as-delivered Vulcan Foundry condition with tall chimney and prominent top feed pipes.

led to a reshuffling of engines to those sheds which had large amounts of freight traffic such as Edge Hill and Aston, although they also took their share of work with the growing number of military specials.

An unexpected sojourn on the Southern Region happened in May 1953 when that region had to withdraw its 'Merchant Navy' Pacifics with axle problems, and then all of its 'West Country/Battle of Britain' Pacifics for examination for the same defects. Amongst the substitutes hastily drafted in for the light Pacifics were seven Black 5s, including 45051 and 45061. They were fitted with SR-style lamp brackets; an extra lamp iron on the right-hand side (as viewed from the front) of the smokebox door between the hinge straps; the footsteps had to be moved in an inch or two and the injector overflow pipes cut back to fit within the tighter loading gauge to keep them away from the electrified third rail conductor rails.

Testing

The LMS wasted no time in testing its new 4-6-0 and 5020 spent three days in September 1934 with the dynamometer car on express passenger trains between Crewe and Euston. It was found that: *the performance of 5020 on these workings was very satisfactory and the work was performed economically. Notable features of the tests were the power development at speed and apparent low running resistance of the engine, plus its steadiness and.*

freedom in running at high coasting speeds. A few days later 5020 carried out similar tests on Manchester-London fitted freights. Steaming of the boiler was *uniform and satisfactory throughout and the engine worked these fitted freight trains very economically.* The report concluded: *From the results above and those of the previous test with this engine on Passenger working, the 2-cylinder 4-6-0 engine appears to be a very efficient and satisfactory mixed traffic power unit.*

Whether the results were viewed through rose-tinted spectacles or the results from the dynamometer car were faulty is not known, but further trials the following month with 5036 on London-Carlisle fitted freights told a different story: *the operation of the engine did not seem to have been satisfactory in respect of the proper use of the regulator in conjunction with the valve gear and some time was lost for this reason.* These poor results were confirmed in March 1935 when 5048, tested on Sheffield-Carlisle fitted freights against two Crab 2-6-0s, which both had 24-element superheaters, came off worst by a worrying margin. The next month further trials were conducted, this time to evaluate the performance of the upgraded superheater arrangement. 5067 with the 14 element (2 Row) superheater was measured against the 21-element (3 Row) 5079 on Carlisle-Sheffield fitted goods trains and St.Pancras-Leeds express passenger workings. The outcome was *a clear reduction in coal and water consumption...*

shown by the 3 Row Element Engine 5079 when compared with the 2 Row Engine 5067, producing worthwhile fuel savings in lbs per DBHP of 18.6% coal and 20.6% water.

The end of steam and preservation

In 1968 when BR steam finally expired, the Black 5s were at the centre of the enthusiast activities in the North West of England. On Sunday 4 August six specials toured Lancashire from Manchester Victoria to commemorate the final day of steam hauled passenger trains. They were hauled by 13 different locomotives over various parts of their journeys, ten of them Stanier Black 5s; the others were two Stanier 8F 2-8-0s, a 'Britannia' 4-6-2 and a BR Class 5. These included two of the original Vulcan Foundry batch, with the second section of the SLS train from Birmingham New Street hauled from Manchester Victoria by 44874 piloting 45017 while 45025 piloted by 45390 worked the LCGB 'Last Day of Steam' special back from Carnforth via Blackburn to Liverpool. Two of the early members of the class were saved for preservation: 5025, one of the unconverted Vulcan Foundry products, and the first Crewe-built engine 5000, which became part of the National Collection.

Black 5s were used on many different types of work. Here is a Western Division example with 5058 passing over Brock water troughs on the LNWR line north of Preston with an up military special. It was allocated at the time to Carlisle Upperby and had received a sloping throatplate boiler in November 1937.

From mid-May to mid-June 1953, seven Black 5s were on loan to the Southern Region whilst axle problems on the Bulleid Pacifics were being put right. 45051 from Monument Lane is ready to depart from Waterloo with the 12.54pm to Salisbury on 22 May. The restricted loading gauge meant that the footsteps had to be moved in by two inches and the injector overflow pipes shortened. Southern style lamp irons with an extra one on the left-hand side of the smokebox door between the hinge straps were fitted to display the SR headcode discs. Photograph R.C Riley, www.transporttreasury.co.uk

One of the last places the Black 5s conquered was the ex-Somerset & Dorset Joint where weight restrictions over several bridges between Mangotsfield and Bath prevented their use over the line until remedial work had been completed in October 1937. Although several were allocated to the line, mostly they were visitors on excursion trains from the north and midlands. 45060 from Crewe South passes the SDJR shed at Bath on 6 July 1957. Photograph L.G. Marshall.

A superb 1950s typical Black 5 with a complete train of carmine and cream coaches, 45065 at Betley Road south of Crewe on the West Coast mainline on 31 March 1956. It was shedded at Aston throughout the decade, was always domeless except for a couple of years pre-war and had a welded tender from 1945. Photograph N.E. Preedy.

45041 holds the gaze of a very young enthusiast at St Helens in late 1964. The Bank Hall engine is on a WCML freight which was bread and butter work for the class for over thirty years. The riveted tender was not recorded on the History Card. Photograph www.rail-online.co.uk

The Black 5s played a key role in the 1968 end of steam activities in the north west. This is one of six specials on 4 August to commemorate the final day of steam hauled passenger train and is the second section of the SLS train from Birmingham New Street departing from Manchester Victoria with 44874 piloting 45017. Photograph www.rail-online.co.uk

45025 with a down fitted freight at Preston on 23 June 1968. By this date the class worked almost entirely on freight duties with just a handful of scheduled passenger turns each week. On Sunday 4 August 45025 worked one of six specials which toured Lancashire from Manchester Victoria to commemorate the final day of steam hauled passenger trains. It was piloted by 45390 on the LCGB 'Last Day of Steam' special from Carnforth via Blackburn to Liverpool. Photograph www.transporttreasury.co.uk

These pictures of 5047 over four decades are a delight for the engine picker with changes in all manner of details every few years as the class was updated and subject to the vagaries of the operating conditions over that time. They also emphasise the dangers of relying on the Record Cards because many of the details are not recorded on the surviving cards and some can only be gleaned by using a magnifying glass and educated guesswork on pictures such as these.

5047 during its first year in service when it was allocated to 12A Carlisle Kingmoor from November 1934 to November 1935. All the original Vulcan Foundry features are visible including tall chimney, prominent top feed pipes, close-spaced LMS on the tender and plain axlebox covers. The crosshead vacuum pump and the plain cylinder wrappers were common to all of the first 75 Black 5s. The livery was standard LMS black lined with vermillion and the insignia and smokebox numberplate were the serif pattern.

3. THE DEVIL IN THE DETAIL

The Black 5s were ordered in quantity straight from the drawing board and it is not surprising that a certain amount of 'tweaking' of the design occurred as they settled down into service. Many of the changes were very minor but others were more fundamental and reflected problems encountered with several of the new Stanier designs; a third group happened over time as the natural result of changing operational and maintenance conditions. To add even more variety, works visits resulted in exchanges of tenders, boilers and later even frames as part of the normal repair processes in an effort to minimise the time spent in the shops.

Boilers

The ability of the boiler to produce sufficient steam when required is a major determinant of the success or otherwise of a steam locomotive and although the early Black 5s did not suffer the same problems as their 'Jubilee' contemporaries, design changes were made before all of the second batch had been completed. It was realised that the relatively low level of superheat brought by Stanier from Swindon was inadequate in a different operating environment and the domeless 14-element superheater boilers fitted to 5020-5069 and 5000-5006 were redesigned to accommodate 21 elements, and these were used on 5007-5019 and 5070

onwards. The only difference in external appearance was the addition of two washout inspection doors with small domed covers on each shoulder of the firebox on 5010 onwards. When boilers were exchanged and an original 14-element boiler replaced a later type blanking plates had to be fitted over the holes left by the doors in the firebox clothing.

Formal recognition of the need to increase the level of superheat came in the approval, in May 1935, of £63,831. 'In order to obtain reduced maintenance costs' it was recorded somewhat ruefully, '167 new Princess Royal, Jubilee and mixed traffic locos were built with a reduced degree of superheat, but work experience has shown that this went too far and a higher degree of superheating will result in an increased operating economy, which will more than meet the additional maintenance cost of improved superheating'. On the Black 5s the rebuilding increased the superheating from 14 to 24 elements, and at the same time problems experienced with the smokebox mounted regulators were addressed by fitting dome mounted regulators, positioned just forward of the firebox. The top feed covers were also modified and the dome-like central portion was replaced by a transverse fairing and since the smokebox regulator lubricator was no longer needed, the streamlined cover

over the atomiser on the smokebox side was reduced in size. The work was done during heavy repairs between March 1937 and October 1940 leaving the 21-element boilers as the only ones to remain domeless.

Under the LMS system of progressive repairs, a pool of spare boilers was created to allow locomotives requiring boiler repairs to be returned to service without waiting for the original boiler to be repaired, which could take much longer than the rest of the engine. When more Black 5s were built in 1936/37 they had 24-element domed boilers and also a larger grate and sloping throatplate to improve combustion, and naturally the spare boilers built as the first heavy general repairs became due on the class were of this design. It was therefore necessary to modify some of the early locomotives with vertical throatplate boilers to accept the sloping throatplate type. In all 18 of the engines covered in this volume were converted; eleven in 1936/37, two in the 1940s, two in the 1950s, and three in 1960; of these six were reconverted to the vertical throatplate type including one which only lasted until its next boiler change.

The final changes in boiler design came in 1947 when the top feed arrangement was changed to reduce the build-up of scaling on the superheater flues. This allowed the top feed to be moved further forward to the first ring

When next pictured during 1937, 5047 had become one of the twelve early conversions to sloping throatplate boiler, in December 1936, when it was repainted in the sans serif 1936 livery, still with 5P5F power classification below the cab side windows. It has retained the tall chimney and crosshead pump but now has access covers in the cylinder wrappers and gutters on the cab roof. It was shedded at 5A Crewe North until June 1937 when it was transferred to Blackpool.

The gloss black has long gone and 5047 has had a wartime repaint with what appear to be hand-painted serif style numbers. It gained its first welded tender in 1946. 5047 has a holder for a single line train staff fitted to the cab side about a foot below the windows.

of the barrel away from the dome, and from 1948 they had a raised fairing over the main cover to accommodate a further modification to eliminate problems with the valves. Only Scottish Region 45011 of the first 70 locomotives received this type of boiler, in 1949 when it went to Crewe for a frame change, reverting to a vertical throatplate boiler in 1954.

The domeless boilers had dome-like top feed covers whereas the domed boilers had transverse covers. The covers were interchangeable and they were sometimes swapped around. This seems to have occurred more often in Scotland where St.Rollox produced

Conversions to sloping throatplate

Engine	Top feed on 2nd ring	Top feed on 1st ring
5002	Dec 37 – W	
5007	Jan 60 – W	
5008	Jan 60 – W	
5011		Jan 49 – Oct 54
5020	Mar 37 – W	
5022	Nov 36 - Oct 58	
5023	Feb 37 - Mar 53	
5026	Feb 37 - Jan 59	
5027	Dec 36 – W	
5040	Nov 36 – W	
5045	Nov 54 – W	
5047	Dec 36 - Sep 55	
5049	Jul 54 - Aug 59	
5054	Jan 37 – W	
5057	Dec 37 – W	
5058	Nov 37 - W	
5059	Jul 45 – W	
5066	Jun 60 – W	

numerous combinations of dome and top feed covers, including some examples having both the separate top feed cover and the cover intended for the dome giving the effect of a double domed engine.

Chimney and Top Feed Pipes
The first fifty Vulcan Foundry engines originally had chimneys 12ft 10½in above rail whereas those built at Crewe, and all subsequent engines, had chimneys which were 2½ inches shorter. 5020-5069 received the latter at their first general repairs. They also originally had the delivery pipes to the top feed clacks outside the boiler clothing under separate prominent covers while on 5000-5019 and 5070-5074 the pipes were under cover strips which were flush with the clothing panels. 5020-5069 soon came into line, probably when their boilers were rebuilt with 24-element superheaters.

Steam Pipe Casings
The first ten Vulcan Foundry engines, 5020-5029, had a 'scalloped' shape cut-out at the lower front of the steam pipe casings where they joined the running plate instead of the very small cut-out on all subsequent construction. Although most of them remained with their original locomotives, some found their way on to other class members following works visits.

Smokebox Door
From 5225 onwards a counter weight

was fitted to the left hand side of the smokebox door/ring to help carry the weight of the door when closed. As boilers and smoke boxes were changed, these could also be seen on the earlier locomotives.

Frames
In order to keep the weight down and achieve maximum route availability the frames were of lightweight construction, only 1 inch thick and lightly stayed, and after a few years in service this came back to bite the LMS as extensive cracks developed, especially at the top corners of the horn gaps, which then had to have sections cut-out, and new pieces welded in and trimmed to shape. As the problem got worse a spare set of frames of the later type designed to accept sloping throatplate boilers was made at Crewe in 1943 to allow engines to be returned more quickly from repair. When the first locomotive requiring major frame repairs arrived, it was stripped and rebuilt on the new frames and once these frames had been repaired they became the spare set. This resulted in many inter-changes of frames, both between the vertical throatplate engines and the later sloping throatplate locomotives, and between locomotives from each builder. Since the builder's plates normally stayed with the frames, this could result in engines that had been built at Crewe or Vulcan Foundry, appearing with Armstrong Whitworth plates and vice versa. The original styles and position of the plates

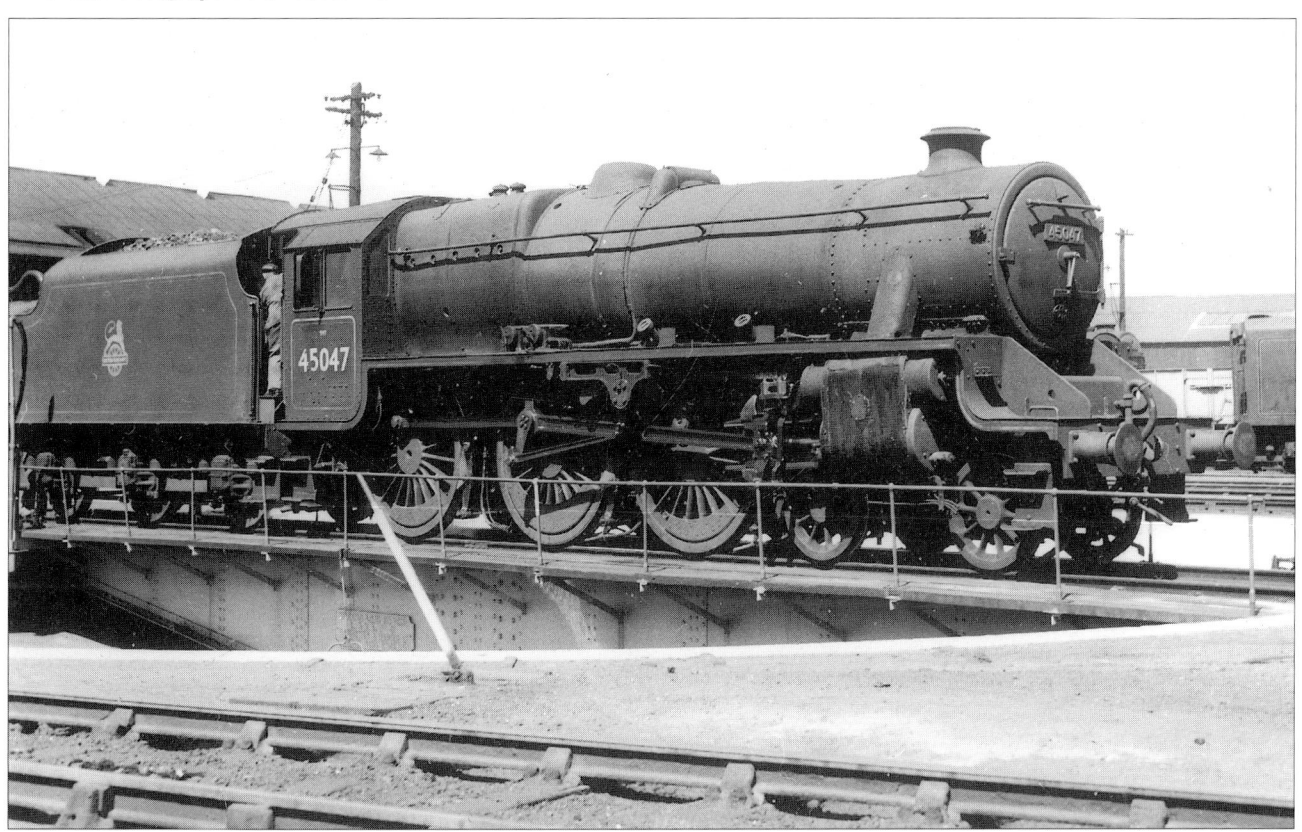

45047, renumbered in August 1948, still with a sloping throatplate boiler but now has a part welded tender with external sieve boxes when photographed at Carlisle Kingmoor in the early 1950s. The livery is BR lined black with the first crest, St Rollox style 10in cab numbers and 1946 pattern smokebox numberplate. The power classification has changed to '5' and the steam lance cock is in the low position.

At first glance this 45047 (at Edinburgh Haymarket) looks identical to the previous picture but closer study reveals that it now has a domed 24 element vertical throatplate boiler which it received in September 1955. This can be identified by the five instead of six washout plugs on the firebox side and the blanking plates on the firebox shoulders where the domed covers had been on the sloping throatplate boiler. It still has the 10in cab numbers but the power classification has changed to 5MT. The tender has also changed, to a welded pattern acquired in 1954, and the steam lance cock is now higher, just below the handrail. Photograph R.K. Blencowe.

In the final picture, taken at Balornock in June 1963, 45047 now has a domeless 21 element boiler as indicated by the domed covers on the firebox shoulders. As was its want, St.Rollox has fitted a topfeed cover from a sloping throatplate boiler. It still has a welded tender and has gained AWS equipment. The livery details have changed with the later BR crest, 8in cab numbers offset to the left to clear the tablet holder and OHLW plates.

The welded variant of the Stanier 4000 gallon tender first appeared with the Armstrong Whitworth 1935 batch of Class 5s. 45044, pictured at Ashton on the West Coast main line with a down freight on 7 March 1963, swapped its original riveted tender for Its first welded tender no.9236 in April 1953. Photograph www.rail-online.co.uk

The part welded tender paired with 45024 on 26 August 1964 at Wigan was fitted at an unknown date. They were introduced in 1944 because problems were experienced with the seams of the welded tanks and found their way onto a small number of the early Class 5s over the next twenty years. Features to note are the rivet pattern on the tank sides and the rectangular vents which replaced the 'mushroom' shaped type on the earlier tenders. Photograph www.rail-online.co.uk

differed between the batches: Crewe-built locomotives had oval plates fitted to the front framing at both sides; Vulcan Foundry 5020-5069 and 5075-5106 had the same pattern but fitted on both sides of the smokebox directly above the top of the steam pipe and above the ejector pipe, although from 5107 onwards they were also attached to the front framing. The first few engines built by Armstrong Whitworth had rectangular pattern plates on the smokebox in the same position as the Vulcan Foundry locomotives and on the remainder the plates were attached to the front framing. The Vulcan plates included the works numbers which were 4565-4614 for 5020-5069 respectively; they were omitted from the standard Crewe plates on 5000-5019 and 5070-5074 although numbers 216-240 were allocated to these engines.

Wheels and Axles

To save weight, the coupled wheels had a 3 inch hollow bore through the axles and the bogie wheels on the first fifty Vulcan Foundry locomotives (5020-5069), and the first four from Crewe (5000-5003) also had hollow axles; all the other engines had a small turning centre machined in the solid axle. The wheels on the first seventy locomotives (5000-5069) and also 1935 Vulcan Foundry 5075-5094 had stiffening webs at the rear of the four spokes adjacent to the

crankpin. Over the years wheelsets were swapped during works visits and all sorts of combinations resulted.

Cylinders and Inspection Covers

5000-5069 had completely plain cylinder lagging sheets with no access holes or covers. A small circular cover plate with four bolt fixing, to provide access to the steam chest drain pipe instead of removing the complete cylinder lagging sheet, was introduced on 5070-5074. In the late 1930s/early 1940s larger covers were fitted and all locomotives eventually carried the later pattern. Two rectangular covers at the top of the lagging sheets were fitted to 5225 onwards and both types of cover plate were gradually fitted to the earlier locomotives.

Combination Levers

5000-5069 had the Horwich design of combination lever which was plain rectangular in section, offset below the spindle guide and forked at the lower end. On 5070-5074 the combination lever was straight and fluted with a slight offset, still forked at the bottom pin, allowing use of the same union link and crosshead arm secured to the crosshead by two bolts.

Running Plate and Cab

The Crewe-built engines, and Vulcan Foundry from 5045 onwards, were built

with a raised step between the frames in front of the smokebox saddle which allowed easier access to the smokebox and top lamp bracket by footplate and shed staff. The first 25 Vulcan Foundry engines didn't have these steps when built but were fitted with them by the late 1930s. Vulcan locomotives could also be identified by the two round-head rivets at each end of the buffer beam; those from Crewe had flush rivets. In BR days round-head rivets were generally used on repairs. 5020-5069 were built without rain gutters at the edges of the cab roof but as the enginemen tired of water running off the roof below the strips and blowing in through the windows or running down over the their heads when leaning out of the cab, gutters were fitted from early 1937; all the others had gutters from new.

Carriage Warming Pipes

Crewe-built engines had a front steam heating valve and pipe fixed to the bottom edge of the buffer beam just to the right of the vacuum pipe dummy but those from Vulcan Foundry did not, although in some cases these appeared later, probably as a result of frame changes. The flexible hoses were usually taken off during the summer months when not needed, and were sent into works for examination, pressure testing and renewal or repair as necessary.

27

LMS black livery lined with vermillion. Vulcan Foundry applied the lining around all four edges of the cab side sheet and below the cab windows, whereas on the remainder the lining was carried straight to the cab roof. The serif insignia were gold leaf shaded red to the right and lake below; the cab numbers were 12in, with the power classification (5P with 5F below) in 3in numerals and a scroll pattern front numberplate. The 15in LMS letters on the tender were spaced at 40in on the Vulcan Foundry engines. The picture also shows plain axlebox covers, hollow axles on coupled wheels, prominent top feed pipes and oval Vulcan Foundry worksplate above steam pipe.

Tablet Exchange Apparatus

Members of the class used on the Northern Division single lines had to have automatic tablet exchange apparatus which was attached to the rear left hand cabside at about footplate height. When not in use, the jaws were held upright against the side plate and as the tablet was to be exchanged they would be swung downward through 90 degrees to project the required distance from the engine to make the exchange. Engines working on other single line routes that used train staffs had holders fitted to the cab sides about a foot below the windows.

Snowploughs

As the primary motive power in the Highlands for around thirty years many of the Scottish-based locomotives were equipped in the winter with small snowploughs fixed to the front bufferbeam with two heavy steel angle uprights held by three fixing bolts about a foot in board from the buffers. With these the engines could run through small drifts whilst hauling normal trains, or as patrolling light engine, preventing the build-up of snow drifts which would otherwise cause a complete blockage if traffic ceased. Most engines fitted with ploughs were based at either Inverness or Carlisle Kingmoor, but a small number allocated to sheds such as Northampton, Springs Branch, Preston and Patricroft were also modified.

BR Days

In addition to the various 'improvements' described in the next chapter the class underwent a number of minor changes in the early 1960s. The two most noticeable were the repositioning of the steam lance equipment and the top lamp bracket. The former was originally low down near the base of the right-hand side of the smokebox, although many were moved just above the top of the steam pipe when boilers were changed and smokeboxes renewed. When the original internal pipes corroded the BR type was fitted which were in the low position with a long external steam feed pipe from above the handrail.

With the onset of electrification from around 1960 'electric overhead' warning flashes, white enamel plates with the symbolic warning sign of forked lightning (in red), were fixed to those parts of the locomotive where footplate crews could come into contact with overhead wires. Also, from late 1963 the upper lamp bracket was for safety reasons moved down to the right of the central door fastening, and the central lamp iron above the bufferbeam was also moved to the right to remain directly under it.

Tenders

Although the initial diagram, ED 177, showed a Fowler pattern tender with 3,500 gallon water and 5½ tons coal capacity, by the time the first Black 5s were built the LMS had decided to equip them with the new Stanier 4,000 gallon coal design carrying 9 tons of coal. The engines which are the subject of this volume were, with three exceptions, all built with these tenders. Crewe-built 5000, 5073 and 5074 were paired with the three prototype 4,000 gallon tenders which had been built in 1933 for the first two 'Princess' 4-6-2s and for the 'Turbomotive' 6202, although this one was actually used for the USA tour of 6100 ROYAL SCOT. Their original flat sidesheets were rebuilt with curved upper sides and they were subtly different from the standard 4,000 gallon tender, with a different rivet pattern and curved cut-out at the top of the side panelling, and two of them had Timken roller bearings, identifiable by their complex axlebox covers. Over the years these three tenders moved around the class, one appearing behind 5002.

The tanks of the original 4,000 gallon tenders were assembled with snap head rivets, but from the first Armstrong Whitworth batch (5125-5224) onwards a change was made to welded tanks, which reduced the weight of the tender by over a ton. However, problems were experienced with the seams and a hybrid type with part welded tanks was introduced in 1944. These two types soon became paired with 5000-5074 because tenders took less time to repair than locomotives and after a works visit an engine would take the next spare available tender, not necessarily the one it arrived with. One other tender was attached briefly to an early Black 5; a

3,500 gallon test tender built in 1937 which ran with 5019.

As with many small details the first fifty Vulcan engines, 5020-5069, differed slightly and their original tenders had plain axlebox covers whereas those from Crewe had cruciform ribs cast in. Also the tank vent pipes on the rear platform behind the coal fender were quite short, and reached only slightly above the level of the tank rear. Soon after the first tenders were built they were extended to the tops of the side plates.

Liveries

With only three exceptions (to be covered in Part 4) the class was always painted black. Insignia and lining-out naturally followed the fashion of the day, with lots of small variations to keep the engine-picker happy.

When built, the engines carried the standard LMS black livery lined with vermillion. Vulcan Foundry applied the lining around all four edges of the cab side sheet and below the cab windows, whereas on the remainder the lining was carried straight to the cab roof. The serif insignia were gold leaf shaded red to the right and lake below; the cab numbers were 12 inch, with the power classification (5P with 5F below) in 3 inch numerals and a scroll pattern front numberplate. The 15 inch LMS letters on the tender were spaced at 40 inches on the Vulcan Foundry engines and 60 inches for those from Crewe.

In February 1936 there was a makeover that brought in fashionable new sans serif insignia to replace the serif characters, and these were applied to all of the 1936/37 built Black 5s from Armstrong Whitworth (which will form the subject of Part 3). The size of the cab numbers was reduced to 10 inches but the letters remained 14 inches high. The change was short-lived and in mid-1937 the LMS reverted to the serif style, albeit with cheaper yellow rather than gold insignia. A number of the earlier locomotives were repainted in the 1936 style including 5002, 5005, 5014, 5023, 5025, 5027, 5029, 5031, 5032, 5036, 5038, 5041-5047, 5051, 5055, 5059 and 5067.

During the war, all full re-paints were plain black although many engines were never fully repainted and probably kept traces of their original lining for many years – even though it was invisible under the layers of grime. The numbers themselves were mostly in 12 inch characters, and the power classification if used at all was abbreviated to 5. After the war plain black continued and most re-paints had scroll and serif characters in yellow (plain or red-shaded), usually with the cab numbers in the 'high' position.

It was not until 1946 that wartime austerity gave way to a new style which had pale straw sans serif characters with inset maroon lining. The lettering was 14 inches and numbers were either 10 inch or 12 inch. Engines known to have received 10 inch numerals were 5015, 5018, 5023 and 5051. Those with 12 inch numerals included 5031, 5043, 5050, 5059 and 5065.

Nationalisation in 1948 brought more livery confusion. Although all the Black 5s repainted in the first part of the year were plain black, there were many variations of insignia. While the new British Railways was deliberating, an M-prefix was applied to the LMS numbers for a short time, although even this was inconsistent. 5048 had 12 inch numbers with a 6 inch high M above and a small figure 5 below. The M was added to the smokebox plate on 5048 with a small extension piece riveted to the end ahead of the numbers. Others including 5064 had the M below the cabside number and the power classification above. Northern Division engines 5012 and 5018 had the M added ahead of their existing numbers. All had BRITISH RAILWAYS on the tender sides in 8 inch cream Gill Sans letters and all retained their LMS smokebox number plates.

The M prefix was dropped by mid-March 1948 and ex-LMS locomotives had 40,000 added to their numbers. Many engines remained for a time in plain black livery. Smokebox door number plates varied between scroll and serif, Gill Sans or 1946 style, and some engines ran for a time with no plates. Cab side numbers were in 10 inch 1946 style characters, 8 inch Gill Sans or 10 inch Gill Sans in various positions and different spacings with power classification figures also in differing positions. A few had a small 4 added in front of their existing LMS numerals. Some tenders still had serif L M S on their sides, others had BRITISH RAILWAYS in 8 inch or 10 inch Gill Sans characters, and a few were devoid of any insignia.

At its home shed of Derby 5031 in the 1936 livery received when it was given a domed boiler during a Heavy General repair in July 1937. Note the BTH speed indicator fitted in February 1938 and the absence of the crosshead driven vacuum pump which was removed in April 1939.

Before the final BR scheme for mixed traffic locomotives was chosen four late-built Black 5s were painted during January 1948 in various experimental liveries at Crewe: three were in the greens of the SR, GWR and LNER and the fourth was in black, with grey/cream/red lining in LNWR style. The latter was selected and although Black 5s started to appear in this livery quite soon after its adoption, many retained their immediate post-war plain black with BR insignia applied for several years and it wasn't until the mid-1950s that all were lined out.

The lettering and numerals were cream Gill Sans edged with a narrow black band. Locomotives repainted in England had 8 inch numbers positioned in line with the tender lettering or emblem whereas St.Rollox started using 10 inch numerals, only changing to the 8 inch type in the mid-1950s. The latter mostly were positioned slightly further down so that their lower edges or even centrelines were in line with the main running plate. The power classification 5 was either immediately above or below the numerals in the same style, although when it changed to 5MT, it usually appeared above the numbers. The smokebox door number plates cast at St. Rollox initially had 1946 LMS style characters whilst those produced at Crewe and Horwich were Gill Sans throughout.

Initially the tenders carried BRITISH RAILWAYS in full, but this was replaced by the larger size of early BR 'lion on a wheel' emblem from around August 1949. From 1957, BR crests approved by the College of Arms replaced the emblem, at first with forward facing lions on each side, but after complaints from the College of Arms all lions faced left.

From December 1963 all locomotives receiving a full repaint were to be painted in plain black. However this does seem not to have been applied by every works and engines outshopped from St.Rollox in early 1964 were still being lined out. Engines repaired at Cowlairs had their shed allocations painted in LNER style on the front buffer beams.

Top right. It was not until 1946 that wartime austerity gave way to a new style; pale straw sans serif characters with inset maroon lining. The lettering was 14 inches high and the cab numbers either 10 inch or 12 inch as shown on 5059. It had received a sloping throatplate boiler at a Heavy General overhaul completed in July 1945 and had been transferred to Nottingham in late 1947.

Below. The engines which were renumbered in early 1948 received a variety of livery and insignia styles. 5020 became 45020 in June 1948 and although it has the BR standard serif smokebox numberplate it has non-standard LMS 1946-pattern cab numbers and a welded tender devoid of ownership markings. It was photographed at Brighton having brought in an excursion, a regular duty and destination for the class in the early post-war years. Photograph www.rail-online.co.uk

45017 at Shrewsbury, probably soon after a Heavy Intermediate completed in April 1950. The first standard BR mixed traffic livery with the LNWR style lining suited the Black 5s. Note the scalloped steam pipe casings which were originally fitted to 5020-5029.

45001 at Crewe South, 10 November 1963. The earliest Stanier tenders had riveted tanks but, to save weight, welded tanks were introduced as standard in 1935 beginning with 5125, the first of the Armstrong Whitworth Black 5s. Unfortunately these tanks were prone to leakage and in 1944 a revised design of combined welded and riveted construction was used. 45001 shows the detail on the rear of a riveted tender including the cut-off corners of the bufferbeam, the guard irons, six footsteps, OHL warning flashes and vacuum and steam hoses. The three plates were only on the Crewe-built riveted tenders: the top one is the tender number, in the middle is the builder's plate and at the bottom the water capacity. The contractor-built tenders and the later tenders had two plates, with the builder's plate omitted. Photograph A.W. Battson, www.transporttreasury.co.uk

45002 at Shrewsbury in July 1959. It was converted to a sloping throatplate boiler in December 1937 and by the date of this picture had received most of the final BR changes. AWS was fitted in February 1959, the steam lance cock has external piping and OHL warning flashes have been applied.

4. ON THE RECORD

Health Warning

As pointed out in earlier volumes of this series the LMS/LMR Engine History Cards and Engine Record Cards, while containing much useful and even fascinating information, should be regarded as a guide to what happened to the engines, not an unimpeachable document to be afforded the status of gospel. It seems to be stating the obvious that the Cards only show what was written on them at the time but the temptation to read and interpret too much should be resisted. Even so, the Cards are a marvellous, fascinating, invaluable record of what happened; yet they are often infuriatingly silent on events that we enthusiasts half a century or more later consider of vital interest and importance. They were filled in, by hand, by clerks and naturally enough contain errors of omission (quite a few) and commission (a few).

Dates of leaving and entry to works were of course to some extent nominal and a day or two either side should always be assumed. Worse, the works were not above 'fiddling' dates slightly at the beginning or the end of a month to enhance the monthly figures, either of engines 'in' or engines 'out'. It was thus not entirely unknown for a locomotive to be out on the road with

the figures showing it still in works and vice versa – for a few days at least. As with all BR steam locomotives, the record fades from about 1959-60 as the people involved realised their charges were on the way out. No-one responsible for the Cards bothered to record the last 'seeing out' mileages on the LMR or other Regions where the Black 5s ended up.

Although History Cards for those engines which were withdrawn from the LMR survive at the NRM, the picture is less complete for the remainder, and especially those which succumbed north of the Border. With a handful of exceptions the post-1950 Cards for these no longer exist and so there are no details of annual mileages or boiler changes from that date, although the works visits on the Record Cards do indicate when a boiler would have been changed. For the very small number of engines which served out their time at WR or NER sheds the ink runs out in the late-1950s, usually around 1957.

Sheds...

The same was true of allocations and many shed moves at the end did not find their way onto the History Cards. However the Chief Accountant's Statistics Office at Derby kept going to

the bitter end of BR steam and beyond, and so the LMR shed allocations from 1963 onwards were taken from the weekly Locomotive Stock Book Alteration Lists produced there. Final gaps were filled for the locomotives allocated to the ScR and NER from the Engine Record Cards which were maintained after the History Card entries ended. The shed descriptions used are as written on the Cards and therefore translation is needed for some of the LMS sheds: Carlisle M = Durran Hill, Carlisle N = Kingmoor, Carlisle W = Upperby; Leeds = Holbeck; Sheffield = Millhouses. Transfer dates were for the week ending, although a few ScR moves can only be tied down to the period ending date.

One of the unsolved mysteries hidden away in the History Cards was the storage for several weeks of many Central Division locomotives in the 1930s. It is understandable that some were hastily stored in September 1939 as the country went to war, but it is not obvious why they were not required in 1936 when new engines were still being delivered. Although most of those involved are dealt with in Part 2, Farnley Junction's 5063 was one of them being stored from 7/10/36-8/11/36. Others which did not earn their keep for several

45010 was allocated to Perth between August 1946 and February 1950 which indicates that this picture was taken in late 1949 or early 1950 after the first BR emblem had been applied during a Light Casual repair completed at the end of September 1949. It has a domeless boiler, a welded tender acquired in March 1945 and a Manson tablet catcher. The three holes visible on the bufferbeam are where a snowplough is attached. www.rail-online.co.uk

weeks included 5060 (Huddersfield 17/10/38-22/12/38), 5006 (Huddersfield 27/2/39-3/4/39, 24/4/39-4/5/39, 8/5/39-22/5/39) and 5047 (Blackpool 13/3/39-3/4/39, 17/4/39-8/5/39). A Western Division engine, 5071 of Rugby was also stored for a month in early 1938 before being loaned out to Millhouses.

Repairs and Maintenance

Under the LMS motive power organisation most sheds carried out minor running repairs and adjustments, including boiler wash-outs etc. Jobs which required the engine to return to Works or to one of the larger sheds, such as Rugby, were usually designated under one of the 'Classified Repair' codes. These were either 'Heavy, (H) or 'Light' (L), further sub-divided into 'Casual' (C), 'Intermediate' (I). 'Overhaul' (O), 'Service' (S) or 'General' (G). Occasionally engines were sent to Main Works for other reasons, such as modifications (e.g. fitting of AWS if this did not coincide with a normally programmed visit) and in these cases the code 'NC' (Non-Classified) was used. The other code which appears from time to time on the Engine Repair cards is 'TRO', which stands for 'Tender Repair Only'. Suffixes, usually after 'NC' were '(EO)', which signified 'Engine Only' and 'Rect.', or 'Rect. (EO)' which was used when an engine had to be returned to Works soon after a works visit for 'rectification'; that is, tightening up bits that had come loose and loosening bits that were too tight.

According to the classification prescribed by the Board of Trade a heavy repair was any one during which an engine was reboilered or had its boiler removed from the frames. It was also when any two of the following were carried out:
Fitting new tyres to four or more wheels.
Fitting new cylinders.
Fitting new axles.
Re-tubing or otherwise repairing the boiler whilst still in the frames with not less than fifty firebox stays renewed.
Both turning wheels and refitting axleboxes.
Stripping and renewing both motion and brake gear.

Intriguingly, light repairs often involved major work such as fitting new axles, replacing cylinders, partially retubing or patching the boiler in situ, or refurbishing the motion, axleboxes, frames, etc. As long as only one of these items was involved, however, the repair was still regarded as light. Most heavy repairs were 'generals' whilst most light repairs were 'intermediates'. General repairs were carried out either at set time intervals or at predetermined mileage windows beyond which it was deemed that an engine could not safely remain in service and were designed to return it virtually to 'as new' condition. Intermediate repairs were normally undertaken when some major component reached the stage where it had to be attended to before the engine was due for general repair, but the aim was to carry out as few intermediate repairs as possible. Thus HG repairs were usually done at approximately 3 to 4-yearly intervals with, typically, two Intermediate repairs (either Light or Heavy) between. The opportunity was often taken to carry out modifications at the Heavy Works visits, though particular programmes, such as installation of ATC/AWS in the late 1950s and early 1960s sometimes required engines to be called in specially.

Works…

Visits to works followed a typical railway ritual: the owning shed would submit a Shopping Proposal, usually some time before it was expected an engine would achieve a mileage or condition beyond which it would be uneconomical or unsafe for it to continue in service. What happened next depended on what was said on the form; either the engine would be called in by the Regional Shopping Control Office, or a Mechanical Inspector would be sent to verify its condition and make his recommendations. The Control Office had to balance between maintaining works loading and keeping the number of engines under or awaiting repair within budgeted targets. This was one of the reasons for the figures being fiddled from time to time.

The LMS Engine History Cards did not record which particular workshop carried out the repairs, although the BR cards and the Engine Record Cards do show this information. The latter dutifully carried on right to the bitter end and so for most locomotives there is a complete list of works visits from cradle to grave. One regional variation which the studious reader will quickly observe is that the Scots appear to have recorded all the minor works visits for rectification much more rigorously than

5018 in wartime plain black in 1948 with an 'M' added ahead of its existing high 10 inch 1946-pattern LMS numbers, and **BRITISH RAILWAYS** in full on its welded tender; the BR number was not applied until March 1950. It is domeless and sports an unusual top feed cover with no separate clack fairings. Photograph W. Hermiston, www.transporttreasury.co.uk

An early and much cared for arrival on the Highland Section, an almost-*burnished* 5022 in 1934. Edge Hill, its next shed, would not look after it like this! 5022 had been delivered from Vulcan Foundry in August – note the open front footplate and scalloped steam pipe casings. It has no shed plate but already has a tablet catcher for working on the single lines in the Northern Division. The wooden building is the old Highland Perth shed, closed a year or two later when the big modern LMS Motive Power Depot opened. Photograph www.transporttreasury.co.uk

The driver of 45066 poses while his fireman looks after the tender of 45066 at Helmsdale on 2 May 1957 with 44798 on the 3.35pm Wick-Inverness. 45066 was allocated to Inverness from May 1942 until February 1960 and gained the welded tender pictured here in January 1957. It was one of the late conversions to sloping throatplate configuration, in June 1960. Photograph Hugh Ballantyne.

their English counterparts. Maybe their engines broke down more often or perhaps this type of attention was given in England at the sheds instead.

Crewe Works was responsible for heavy repairs to Black 5s on the LMS Central, Midland and Western Divisions, and St.Rollox (Glasgow) for those allocated to the Northern Division, although it was noted in 1942 that in most weeks two of its allocation of Black 5s and Jubilees were being sent to Crewe for these repairs and this continued until the mid-1950s.

An HG repair at Crewe in the LMS period usually took between 25 and 40 Weekdays: this was insufficient time to do the necessary work on the boiler, so engines undergoing HG Repairs invariably left Works with a different boiler from that with which they had entered. Matters continued after 1948 in a similar vein until about 1962 with Crewe maintaining the LM Region engines and any others allocated at various times to the Southern Region (S&DJ), Western Region (mainly Shrewsbury) and the North Eastern Region (after transfer of the Yorkshire sheds). St.Rollox looked after the Scottish engines, including those at Carlisle Kingmoor. From around 1963 St.Rollox and Crewe started to run down their steam activities at least as far as Black 5s were concerned and Cowlairs began to repair Black 5s from both Regions. After Cowlairs closed in 1966, Crewe did a few final Heavy repairs to LM engines; by this time repairs to Scottish engines seem to have ceased.

Mileages
Total annual mileage was recorded on the History Cards, but only up to around 1960 and, because the Cards have been lost, is not available for most of the ScR engines after 1950. In any event it was an estimation, a minor miracle of paperwork and not mechanically recorded. Having said that, a few snippets can be extracted to show how the early Black 5s compared with the engines they displaced.

In 1936, the first year in which all the engines covered in this volume were in service for a full twelve months, the average mileage was 51,377, with Scotland's 5016 the highest at 67,731 and Crewe's 5000 letting the side down a bit with only 35,304. The overall figures compared favourably with 20-28,000 miles for the LNWR 'Precursor' and 'George V' 4-4-0s and the Highland 4-6-0s, 33,000 miles for the Caledonian 4-6-0s and 37,000 for the LNWR Prince of Wales. The new Black 5s were generally not affected by regular works visits in their first twelve months or so in service, and these would account for at least one month in each year as the class fell into the normal ongoing repairs regime.

As more of the engines came into service and the class worked lesser duties as well as the longer distance services on which they were originally used, the average mileages reduced. The cumulatives to 31 December 1950, the last date up to which figures survive for every engine, show a fall to an average of around 39,500 per annum. The Scottish contingent were generally

worked harder than many of their English counterparts and the star performer was 5010 which reached 815,208 miles, an annual average of 51,000. 5007-9, 5011-18, 5023, 5029, 5036, 5053 and 5066 which all spent most of their time on the Northern Division and 5065 and 5068, also achieved around 750,000 miles in the same period. The wooden spoon went to 5038 which was passed around between different Western Division sheds, being reallocated no less than twelve times, and managed only 532,378 miles during the 16 years.

'Improvements, Etc'
The History Cards had a section headed 'Improvements, Etc.' that recorded brief details of modifications or improvements applied to the engines. What was recorded here varied from major work such as the fitting of speed indicators or AWS right down to apparently trivial jobs costing a few pounds. The clerks responsible for the cards did not always record changes which must have taken place such as the removal of the crosshead vacuum pumps which was done on all the pre-war engines, nor were they necessarily consistent with the descriptions and Works Order numbers used as a shorthand for this purpose. That said, it is possible with a little detective work to get a reasonably accurate picture of what was done, when. This book uses the relevant 'off works' date for each particular modification rather than the 'period ending' date which was actually written on the Improvements section of

5018 in the late 1930s, probably soon after a light repair in May 1938 during which its crosshead vacuum pump was removed. It still has it original domeless 21-element boiler with two washout inspection doors under small domed covers on each shoulder of the firebox. It was allocated to Inverness at this date, hence the tablet catcher which is obscuring the final digit of the cab number; the numbers were moved out of the way from the 1940s onwards. Photograph M. Robertson, www.transporttreasury.co.uk

A soaking at Stoke. Photograph J.M. Vaughan, www.transporttreasury.co.uk

the cards because clearly this makes more sense.

Tabulated in the following pages for each engine are the most significant 'Improvements' together with boiler and tender changes. The main boiler types, as discussed in the previous chapter, are indicated on the individual engine histories which follow, as are changes in tender type. Where a boiler or tender change is known to have occurred or a modification applied but the date is not recorded a '?' is shown in the tables. To help interpretation, if a change from the as-built type occurred then this has been indicated, e.g. a locomotive built with a domeless boiler and riveted tender receiving a domed boiler or a welded tender. Note that the History Cards stopped recording boiler changes before the end, when a locomotive clearly may have had at least one more boiler after that shown.

Crosshead driven vacuum pumps

The crosshead pump, which was attached to the bottom of the lower left-hand slide bar, was intended to maintain the vacuum in the train pipe when running. Enginemen were instructed to use the small ejector to maintain the vacuum when standing in a station, or just before starting, but as soon as the engine was running fast enough for the pump to maintain the vacuum, the small ejector should be shut off to conserve steam. In service, the pumps proved to be unreliable and costly to maintain and were not used, the crews preferring to use the small ejectors to maintain the vacuum, and they were removed between 1938 and 1941.

Steam Sanding

Sand was originally applied to the rails by gravity from six sandboxes supplying the six coupled wheels but this method proved unsuccessful and by 1938 steam sanding was introduced. It is uncertain when this work was actually carried out on some of the Northern Division engines because many of their History Cards show it dated either 19/5/45 or 20/5/45, obviously a purely housekeeping exercise with the modification actually done some years earlier.

Speed Indicators

The LMS experimented with speed indicators and recorders on a number of classes right up to the time of nationalisation, the M&EE Committee approving in October 1937 the recommendation that, 'with a view to enabling engines accurately to observe speed restrictions… 998 locos working express passenger trains be fitted with electric speed indicators'. Between 1938 and 1943 a number of Black 5s were fitted with British Thomson Houston speed recorders which had the alternator mounted on a bracket suspended from the running plate alongside the left-hand trailing wheel. It was driven by a pin on the end of a small return crank from the driving wheel crankpin with a voltmeter in the cab graduated to indicate the speed. The equipment proved unreliable and, with wartime, spares were in short supply, and its removal was ordered in 1944.

Further trials with modified BTH and Smith-Stone speed indicators began in 1949 but slow progress was made until

1957, when the ex-LMS Pacifics were fitted with the latter type. In May 1959 authority was given for the widespread fitting of this equipment and though it was decided that initially 99 Black 5s would receive it, the programme was cancelled in 1964 with only a few of the class equipped. These indicators were again electrically operated, speed being calculated from the voltage produced by a generator. This was mounted directly on and driven from a return crank on the rear left-hand crankpin, with an armoured flexible cable leading via a rheostat box into the cab. Why the term 'return crank' was used is a mystery since there was no crank action; the arm simply rotated a pin concentric with the driving wheel centre.

Modification and Modernisation

The long-running problem of the frame fractures caused much head-scratching and prompted several abortive solutions before it was finally solved in the 1950s. In December 1938 the M&EE Committee heard that, 'During the course of repairs to Black 5 4-6-0 MT engines, it had been found that several of the frames had been bent inwards near the leading axlebox and to eliminate this trouble he [Stanier] recommended that cross stretchers be fitted in the region of the axlebox guides to the 472 engines concerned'. Unfortunately this only seemed to increase the incidence of cracks, and the cause was eventually traced to the transmission of the racking stresses between the frames which resulted in further loosening of the guides and stays. Attempts to repair the frames by chipping away vee-shaped

45022 from Dalry Road shed, at Beattock on 22 August 1952 with large size cab numbers positioned low down to clear the tablet holder and BRITISH RAILWAYS in full on the tender, a legacy of its early renumbering in July 1948 after a Heavy General repair. It had a sloping throatplate boiler from November 1936 until October 1958. Photograph J. Robertson, www.transporttreasury.co.uk

grooves along the cracks with an air caulking gun and then welding them up, were not successful and the cracks duly reappeared at the edges of the welds.

The underlying weakness of the frame design meant that none of these modifications cured the problems, although eventually the cause was beginning to be understood, largely as a result of trials and experiments carried out by the Research Department. In a paper read to the Institution of Locomotive Engineers in 1946, E.S. Cox stated that these investigations had shown the 'overwhelming importance' of a tight connection at the bottom of the horn gap. He went on to say that the types of axlebox guides and hornstays used on the Black 5s prevented this from being achieved. The results bore fruit, however, in the Black 5s built during 1946. These had two features which significantly improved things: firstly manganese steel liners were fitted on both the axleboxes and horns. These hardened in use so that after a short time the rate of wear slowed down giving a major improvement, the clearances and alignment of the boxes with the frames being maintained to within very close limits over much higher mileages than previously experienced. Secondly, adoption of Horwich pattern hornstays as used so effectively for many years on the Crab 2-6-0s. Problems with the screwed spring links were also dealt

with by the use of flat section links with large box-section brackets through which were inserted interchangeable flat cotters, by which means the weight was adjusted on the coupled wheels.

The early engines were therefore fitted with manganese steel liners, Horwich hornstays and cottered links as they underwent general repairs from 1947 onwards, although financial authority was not given until 1951 when Job No.5597 to WO/E 1173 was issued to Modernise (or Modify as some of the History Cards describe the work) the 643 locomotives which had not already been dealt with informally. Although the costs were over £600 per engine the results were certainly impressive and the interval between periodic repairs rose from an average of just under 57,000 to over 97,000 miles. However, with the end of steam approaching not all the engines received the modified spring links, many retaining the screwed type until withdrawn.

AWS

From 1959 onwards most of the class were fitted with the BR Automatic Warning System (AWS), which was also recorded on the History Cards as Automatic Train Control (ATC). The main visible features were a cylindrical vacuum reservoir on the right-hand running plate immediately in front of the cab with a smaller timing reservoir on the left-hand side. An extra frame

stretcher was added to the front of the bogie to which the AWS receiver was fixed, with a guard plate attached to the buffer beam to prevent the screw coupling damaging the receiver.

Above. **The coal on 45049 receives close attention at Haymarket in November 1954. 45049 had been fitted with a sloping throatplate boiler in June 1954 which it kept until August 1959. The welded tender was one of three it was paired with over the years, although it ended up with a part-welded example. Note the St Rollox use of 1946 LMS style characters for their smokebox number plates.**
Photograph J. Robertson,
www.transporttreasury.co.uk

45000

Built as 5000 at Crewe Works 23/2/35
Renumbered 45000 w.e. 22/1/49

Improvements and modifications
15/3/39	Steam sanding
15/3/39	Removal of vacuum pump
6/11/40	BTH speed indicator
4/4/59	Fitting BR ATC equipment

Repairs
12/2/36-28/2/36	LS
5/9/36-1/10/36	LO
22/5/37-8/6/37	HS
20/10/37-29/11/37	LO
20/2/39-15/3/39	HG
23/10/40-6/11/40	LS
20/7/42-15/8/42	TRO
28/12/42-16/1/43	HG
18/5/44-2/6/44	HS
8/11/45-14/12/45	LS
6/7/47-27/8/47	LS
20/12/48-22/1/49	HG
25/1/50-21/2/50	HI
30/8/51-27/10/51	LI
9/3/53-24/4/53	HG
9/5/53-23/5/53	NC(EO Rec)
21/12/53-18/1/54	LC(EO)
21/4/54-10/5/54	LC(EO)
18/11/54-5/1/55	LI
31/10/55-16/11/55	LC
21/7/56-14/8/56	HI
26/8/57-21/9/57	HG
21/10/57-8/11/57	NC(Rec)(EO)
2/3/59-4/4/59	LI
19/9/59-27/11/59	LC(EO)
31/1/61-10/3/61	HI
7/1/63-29/1/63	HI
10/4/64-5/6/64	LI
3/11/64-19/11/64	LC
20/9/65-25/10/65	LI

Boilers
New	8817
15/3/39	9036 from 5106
16/1/43	8957 from 5163
2/6/44	8683 from 5092 (domed)
22/1/49	8685 from 5134 (domed)
24/4/53	8919 from 45096
21/9/57	9030 from 45144

Tenders
New	9002 (built for Engine 6100 for Chicago World Fair)
16/1/43	9252 (welded)
27/8/65	10528
26/10/65	9252

Mileage/(weekdays out of service)
1935	56,390 (18)
1936	35,304 (56)
1937	34,202 (75)
1938	37,768 (31)
1939	33,984 (54)
1940	31,056 (46)
1941	33,675 (32)
1942	26,647 (77)
1943	43,515 (50)
1944	44,351 (65)
1945	40,877 (107)
1946	51,766 (45)
1947	26,711 (103)
1948	34,129 (68)
1949	44,453 (64)
1950	41,624 (62)
1951	27,372 (74)
1952	40,595 (32)
1953	34,821 (82)
1954	30,245 (87)
1955	32,511 (70)
1956	41,998 (63)
1957	28,616 (88)
1958	38,543 (46)
1959	35,158
1960	33,928

Mileage at 12/36: 91,694
Mileage at 31/12/50: 616,452

Sheds
Bristol	23/2/35
Carlisle North	30/3/35
Crewe	5/10/35
Rugby	5/6/43
Speke Junction	26/1/52 (loan)
Rugby	9/2/52
Crewe North	17/9/55
Rugby	24/9/55
Crewe North	3/3/56
Crewe South	27/10/56
Carlisle Upperby	9/2/57
Crewe South	7/12/57
Holyhead	20/6/59
Crewe South	19/9/59
Carnforth	10/6/61
Crewe South	16/9/61
Holyhead	22/6/63
Chester (M)	2/11/63
Lostock Hall	27/5/67

Stored
25/7/66-19/5/67

Withdrawn w.e. 28/10/67

45000 was not the first of the class as many presumed, but the first built at Crewe. It was allocated to Chester when recorded in unlined black at Willesden on 2 July 1964. It reverted back to a domeless boiler in April 1953 after having a domed example for the previous nine years. Its welded tender was acquired in January 1943. AWS was fitted in April 1959 and it has a lowered top lamp bracket and electrification OHW flashes. 45000 was preserved in the National Collection after withdrawal in October 1967. www.rail-online.co.uk

45001

Built as 5001 at Crewe Works 23/2/35
Renumbered 45001 w.e. 19/3/49

Improvements and modifications

22/8/44	Steam sanding
21/2/59	Fitting BR ATC equipment

Repairs

8/4/36-27/4/36	LS
1/4/37-1/5/37	HS
4/2/38-14/3/38	HG
9/3/40-21/3/40	HS
24/7/41-8/8/41	HG
12/12/42-2/1/43	LS
8/8/44-22/8/44	HS
10/5/46-30/5/46	LS
20/6/47-1/8/47	HG
16/2/49-14/3/49	LI
10/2/50-2/3/50	LC
3/10/50-7/11/50	HG
17/11/51-15/12/51	LI
19/10/53-7/11/53	HI
16/7/54-14/8/54	LC(EO)
23/9/55-9/10/55	HG
14/10/56-21/11/56	LI
30/7/57-24/8/57	HC(EO)
27/8/58-3/10/58	LI
16/2/59-21/2/59	NC(EO)
24/2/60-8/4/60	LI
7/8/61-14/9/61	HG
1/1/63-1/11/63	LI
10/5/66-21/6/66	LI

Boilers

New	8818
24/2/38	9020 from 5090
8/8/41	8921 from 5184
22/8/44	8987 from 5041
1/8/47	9029 from 5223
7/11/50	9034 from 5021
19/10/55	9050 from 45150
14/9/61	9008 from 45217

Tenders

New	9054

Mileage/(weekdays out of service)

1935	52,303 (16)
1936	40,268 (23)
1937	33,798 (46)
1938	34,953 (65)
1939	36,766 (48)
1940	29,775 (42)
1941	32,718 (76)
1942	34,395 (38)
1943	40,111 (34)
1944	26,776 (58)
1945	27,705 (66)
1946	26,984 (55)
1947	20,647 (94)
1948	37,311 (51)
1949	37,162 (84)
1950	30,738 (86)
1951	41,358 (54)
1952	41,395 (38)
1953	28,692 (67)
1954	35,676 (63)
1955	30,788 (81)
1956	36,684 (61)
1957	38,559 (59)
1958	32,154 (81)
1959	36,030
1960	33,939

Mileage at 12/36: 92,571
Mileage at 31/12/50: 542,410

Sheds

Bristol	23/2/35
Carlisle North	30/3/35
Crewe	19/10/35
Aston	27/2/43
Warrington	10/4/43
Holyhead	7/7/51
Crewe North	15/9/51
Speke Junction	26/1/52 (loan)
Mold Junction	9/2/52
Crewe South	7/11/59
Crewe North	20/6/64
Crewe South	19/9/64
Rugby	28/11/64
Nuneaton	12/6/65
Holyhead	4/6/66
Carnforth	10/12/66

Withdrawn w.e. 16/3/68

45001 was one of the class which saw little change during its three decades in service. It carried a domeless boiler and retained its original riveted tender throughout. AWS was fitted in February 1959 and it has the usual final modifications of lowered top lamp bracket and OHW flashes. Seen here at Bradford in 1967 it has a Carnforth shedplate following its transfer at the end of 1966, and it remained there until withdrawn in March 1968. www.rail-online.co.uk

45002

Built as 5002 at Crewe Works 27/2/35
Renumbered 45002 w.e. 30/10/48

Improvements and modifications

28/12/37	Sloping throatplate boiler
22/4/39	Removal of vacuum pump
23/6/44	Steam sanding
5/2/59	Fitting BR ATC equipment

Repairs

3/4/36-21/4/36	LS
16/4/37-10/5/37	HS
19/11/37-28/12/37	HG
23/11/39-21/12/39	LS
16/5/41-7/6/41	HG
9/11/42-25/11/42	LS
9/6/44-23/6/44	HG
26/6/45-31/7/45	LS
25/11/45-15/12/45	LO
16/6/46-18/7/46	LS
12/9/47-17/10/47	HS
8/10/48-26/10/48	LO
8/11/48-16/11/48	LO
13/6/49-29/7/49	HG
12/6/50-24/7/50	LI
10/3/52-29/3/52	HI
17/10/53-12/11/53	HG
19/11/53-23/11/53	NC(Rec)
29/5/55-22/7/55	HI
15/2/56-2/3/56	LC(EO)
23/2/57-22/3/57	LI
18/11/58-20/12/58	HG
29/1/59-5/2/59	NC(EO)
31/1/61-8/3/61	LI

Boilers

New	8819
10/12/37	10132 New (sloping throatplate)
7/6/41	9452 from 5332 (sloping throatplate)
23/6/44	9369 from 5328 (sloping throatplate)
29/7/49	9355 from 5394 (sloping throatplate)
12/11/53	9735 from 45442 (sloping throatplate)
20/12/58	9443 from 45252 (sloping throatplate)

Tenders

New	9055
31/7/45	9001
15/12/45	9114

Mileage/(weekdays out of service)

1935	52,742 (15)
1936	39,952 (26)
1937	32,209 (78)
1938	39,160 (23)
1939	36,909 (43)
1940	31,962 (38)
1941	30,064 (43)
1942	30,540 (43)
1943	45,624 (22)
1944	42,274 (60)
1945	35,148 (86)
1946	42,373 (67)
1947	32,797 (81)
1948	36,721 (83)
1949	39,616 (71)
1950	40,424 (61)
1951	36,283 (45)
1952	33,166 (47)
1953	23,271 (66)
1954	39,279 (54)
1955	32,991 (87)
1956	32,070 (78)
1957	38,423 (46)
1958	26,838 (71)
1959	43,391
1960	35,764

Mileage at 12/36: 92,694
Mileage at 31/12/50: 608,515

Sheds

Bristol	27/2/35
Carlisle North	11/5/35
Crewe North	5/10/35
Rugby	5/6/43
Bletchley	26/8/44 (loan)
Rugby	28/6/47
Crewe North	3/3/56
Crewe South	11/3/56
Carlisle Upperby	31/8/57
Crewe South	5/10/57
Holyhead	20/6/59
Crewe South	19/9/59
Carnforth	10/6/61
Preston	15/7/61
Crewe South	9/9/61

Withdrawn w.e. 3/7/65

45002 with a stopping train at Hunton Bridge, south of Kings Langley, on 9 April 1955. It had been shedded at Rugby since June 1947 and always had a riveted tender. 45002 was one of thirteen Black Fives to be converted in the late-1930s to take a sloping throatplate boiler and it ran in this form from December 1937 until withdrawn.

45003

Built as 5003 at Crewe Works 5/3/35
Renumbered 45003 w.e. 16/10/48

Improvements and modifications
30/1/39   Steam sanding
30/1/39   Removal of vacuum pump
24/2/59   Fitting BR ATC equipment

Repairs
19/2/36-6/3/36	LS
8/3/37-25/3/37	HS
3/4/38-9/4/38	LO
30/12/38-30/1/39	HG
22/7/40-5/8/40	LS
16/5/42-30/6/42	HG
6/5/43-29/5/43	LO
24/6/44-15/7/44	LS
14/4/45-3/5/45	LO
7/11/45-22/11/45	HG
18/4/47-26/5/47	LS
28/9/48-15/10/48	LS
23/3/49-29/4/49	HC
9/3/50-4/5/50	HG
9/7/51-2/8/51	HI
22/3/53-17/4/53	LI
1/7/53-22/8/53	LC
23/6/55-5/8/55	HG
23/5/57-14/6/57	HI
23/6/58-12/8/58	LC(EO)
17/2/59-24/2/59	NC(EO)
11/4/59-13/5/59	HI
21/8/60-11/10/60	HG
5/9/61-20/10/61	LC(EO)
19/11/63-24/12/63	LI
20/7/64-21/8/64	LC

Mileage at 12/36: 88,086
Mileage at 31/12/50: 625,473

Sheds
Perth	5/3/35
Crewe North	2/11/35
Monument Lane	4/3/39
Bangor	1/7/39
Llandudno Jcn	30/12/39
Crewe	6/4/40
Edge Hill	15/2/41
Crewe	25/7/42
Rugby	5/6/43
Willesden	31/10/53
Crewe North	14/4/56
Crewe South	20/10/56
Crewe North	5/1/57
Crewe South	23/11/57
Stoke	20/6/59
Crewe North	12/9/59
Stoke	10/6/61

Withdrawn w.e. 27/5/67

Boilers
New	8820
16/1/39	8965 from 5185
20/6/42	8972 from 5033
22/11/45	8661 from 5204 (domed)
4/5/50	8643 from 5061 (domed)
5/8/55	9038 from 45035
11/10/60	8997 from 45197

Tenders
New	9056
13/7/51	10491 (welded)
11/7/64	10532 (welded)

Mileage/(weekdays out of service)
1935	52,753 (32)
1936	35,333 (54)
1937	36,299 (35)
1938	28,605 (43)
1939	41,302 (68)
1940	35,325 (40)
1941	31,847 (46)
1942	26,918 (65)
1943	41,402 (46)
1944	43,724 (51)
1945	33,007 (63)
1946	51,849 (44)
1947	41,147 (69)
1948	43,399 (58)
1949	44,035 (69)
1950	38,528 (71)
1951	38,942 (55)
1952	39,859 (47)
1953	33,808 (85)
1954	40,184 (39)
1955	32,063 (78)
1956	42,370 (42)
1957	42,412 (46)
1958	32,838 (93)
1959	38,249
1960	33,400

45003 at Willesden on 29 August 1964 in unlined livery and with a crude hand-painted 5D Stoke shedcode on the smokebox door. It is domeless having reverted back in August 1955 after carrying a domed example for the previous ten years. It had a welded tender from July 1951 onwards. AWS was fitted in March 1959 and it has a lowered top lamp bracket and electrification OHW flashes.

45004

Built as 5004 at Crewe Works 12/3/35
Renumbered 45004 w.e. 22/5/48

Improvements and modifications
1/8/38	Removal of vacuum pump
11/10/39	Steam sanding
11/10/39	BTH speed indicator
10/1/59	Fitting BR ATC equipment

Repairs
19/2/36-4/3/36	LS
1/6/37-14/6/37	HS
28/6/38-1/8/38	LS
20/9/39-11/10/39	HG
4/9/41-27/9/41	HS
3/3/43-18/3/43	LS
16/10/43-2/11/43	LO
13/10/44-28/10/44	HG
21/6/46-23/7/46	LS
17/12/46-16/1/47	LO
8/4/48-17/5/48	HG
16/4/49-2/7/49	HI
30/8/50-19/9/50	LI
5/4/52-7/6/52	HG
14/2/53-18/3/53	LI
15/9/53-9/10/53	NC
30/12/53-26/1/54	LI
31/8/55-27/9/55	HI
18/6/56-25/7/56	LC(EO)
5/4/57-16/5/57	HG
12/5/58-27/5/58	LC(EO)
10/12/58-10/1/59	HI
22/4/61-26/5/61	HI
25/2/63-23/3/63	HG
24/6/65-31/7/65	LI

Mileage/(weekdays out of service)
Year	Mileage
1935	58,188 (15)
1936	49,400 (62)
1937	47,160 (48)
1938	48,022 (93)
1939	41,946 (68)
1940	39,666 (46)
1941	35,975 (70)
1942	31,137 (27)
1943	29,107 (56)
1944	32,948 (37)
1945	38,816 (49)
1946	37,370 (77)
1947	40,450 (53)
1948	39,631 (67)
1949	43,703 (58)
1950	37,963 (57)
1951	44,591 (36)
1952	44,960 (83)
1953	38,715 (59)
1954	30,928 (46)
1955	38,948 (51)
1956	41,179 (53)
1957	40,656 (73)
1958	39,559 (76)
1959	44,701
1960	43,104

Boilers
New	8821
20/9/39	8923 from 5143
28/10/44	8921 from 5001
17/5/48	9008 from 5099
7/6/52	8673 (domed)
16/5/57	9026 from 45143
?	8639 (domed)

Tenders
New	9057
18/2/45	9627 (welded)

Mileage at 12/36: 107,588
Mileage at 31/12/50: 651,482

Sheds
Perth	18/3/35
Crewe	2/11/35
Stoke	20/6/36
Tebay	3/10/36
Crewe	3/7/37
Stoke	14/8/37
Rugby	25/9/37
Llandudno Junction	15/10/38 (loan)
Carnforth	5/11/38
Llandudno Junction	12/11/38 (loan)
Preston	17/12/38
Edge Hill	27/3/43
Crewe	13/5/44
Rugby	10/6/44
Carnforth	4/11/50
Patricroft	13/6/53
Crewe South	15/8/53
Springs Branch	21/11/53
Bletchley	26/2/55
Longsight	14/6/58
Crewe North	20/9/58
Crewe South	6/1/62
Llandudno Junction	22/6/63

Withdrawn w.e. 1/10/66

Bletchley's 45004 on the Southern Region at Bognor on 30 June 1957, a regular destination for Black Fives on excursions from the North and Midlands. It is still smart after emerging from an HG overhaul the previous month where it gained the final style of BR crest as it reverted back to a domeless boiler, having had a domed type from June 1952. 45004 is paired with a welded tender which it had from February 1945 onwards. Note that it has an Armstrong Whitworth worksplate having received the frames from 45249 in April 1948. Photograph L.G. Marshall.

45005

Built as 5005 at Crewe Works 12/3/35
Renumbered 45005 w.e. 11/9/48

Improvements and modifications

6/9/39	Removal of vacuum pump
6/9/39	BTH speed indicator
12/7/47	Steam sanding
25/1/55	Modernisation
14/3/59	Fitting BR ATC equipment

Repairs

4/3/36-20/3/36	LS
21/4/36-27/5/36	HO
17/5/37-31/5/37	HS
15/11/37-22/12/37	HG
13/7/39-15/7/39	TRO
23/8/39-6/9/39	HS
13/9/40-11/10/40	HG
20/5/41-27/6/41	LS
2/11/42-31/12/42	HS
18/10/43-13/11/43	LS
21/8/44-16/9/44	LS
3/10/45-7/11/45	HG
20/12/46-1/2/47	LS
4/8/48-11/9/48	HS
1/12/48-20/1/49	LC
7/6/49-27/6/49	LC
1/10/49-29/10/49	HG
10/5/51-1/6/51	LI
16/9/52-16/10/52	LI
26/10/53-21/11/53	LI
6/11/54-25/1/55	HG
2/5/56-26/5/56	LC
23/4/57-17/5/57	HG
9/2/59-14/3/59	LI
9/6/61-26/7/61	HI
13/2/63-7/3/63	HC
5/6/64-10/10/64	HI

Boilers

New	8822
6/12/37	8915 from 5135
11/10/40	8676 from 5031 (domed)
7/11/45	8995 from 5155
29/10/49	9054 from 5014
17/5/57	8670 from 45076 (domed)

Tenders

New	9058
13/10/42	9184
3/9/44	9194
18/9/44	9264 (welded)
27/9/44	9184
7/11/45	9592 (welded)
17/9/52	9536 (welded)
7/12/60	9482 (welded)

Mileage/(weekdays out of service)

Year	Mileage (weekdays out of service)
1935	60,613 (22)
1936	54,210 (71)
1937	29,944 (82)
1938	42,541 (35)
1939	41,326 (49)
1940	42,769 (56)
1941	38,586 (71)
1942	53,238 (7)
1943	51,495 (61)
1944	47,302 (74)
1945	38,716 (94)
1946	43,215 (41)
1947	35,481 (124)
1948	26,740 (89)
1949	39,928 (97)
1950	43,321 (26)
1951	34,929 (38)
1952	36,396 (56)
1953	41,490 (54)
1954	31,475 (81)
1955	40,672 (59)
1956	45,798 (44)
1957	50,919 (40)
1958	48,693 (30)
1959	46,136
1960	41,587

Mileage at 12/36: 114,823
Mileage at 31/12/50: 689,425

Sheds

Perth	13/3/35
Crewe	2/11/35
Farnley Junction	4/6/38 (loan)
Huddersfield	18/6/38
Inverness	6/12/41 (loan)
Inverness	27/12/41
Carlisle Kingmoor	11/11/44
Perth	27/7/46
Carlisle Kingmoor	3/5/47
Edge Hill	9/4/49 (loan)
Edge Hill	30/4/49
Llandudno Junction	20/12/52
Holyhead	13/6/53
Patricroft	19/9/53
Edge Hill	3/4/54

Withdrawn w.e. 27/1/68

Below. Snowplough fitted 45005 was renumbered in September 1948 and carries the full **BRITISH RAILWAYS** on its welded tender. It exhibits a few St.Rollox touches including a LMS 1946-pattern smokebox door numberplate and an 'incorrect' type of cover on its domeless boiler. It was allocated to Carlisle Kingmoor at the date of this photograph until it moved to Edge Hill in April 1949.

45006

Built as 5006 at Crewe Works 15/3/35
Renumbered 45006 w.e. 21/5/49

Improvements and modifications

16/4/38	Removal of vacuum pump
1/9/43	Steam sanding
16/9/54	Modernisation
24/9/60	Fitting Smith-Stone speedometer

Repairs

30/3/36-14/4/36	LS
17/2/37-19/3/37	HG
30/12/38-11/2/39	HS
18/6/40-10/7/40	HG
22/9/41-18/10/41	LS
24/8/42-26/9/42	LS
31/7/43-1/9/43	HG
24/7/44-17/8/44	LS
3/10/45-8/11/45	HG
4/9/46-8/10/46	LS
28/1/47-22/2/47	LO
13/5/47-26/5/47	LO
2/12/47-8/1/48	LS
20/2/48-13/3/48	LO
2/7/48-6/7/48	NC
13/4/49-20/5/49	LI
9/8/49-3/9/49	LC
8/9/49-16/9/49	NC(Rec)
4/3/50-5/4/50	HG
27/2/52-17/3/52	HI
12/9/53-2/10/53	LI
29/10/53-13/11/53	NC(Rec)
20/8/54-16/9/54	HG
12/4/56-9/5/56	LI
19/9/57-28/10/57	HG
17/11/58-5/12/58	LC(EO)
18/8/60-24/9/60	LI

Boilers

New	8823
5/3/37	8639 from 5022 (domed)
10/7/40	8927 from 5147
1/9/43	8828 from 5018
8/11/45	9028 from 5167
5/4/50	8829 from 5030
16/9/54	8998 from 45212
28/10/57	8941 from 45071

Tenders

New	9059
27/11/35	9083
18/6/43	9258 (welded)
7/11/45	9184
19/5/47	9083
2/11/56	9589 (welded)
12/9/58	9496 (welded)
11/8/62	10534 (welded)
25/1/64	10757

Mileage/(weekdays out of service)

1935	59,223 (24)
1936	58,502 (51)
1937	35,893 (61)
1938	35,732 (71)
1939	29,360 (116)
1940	34,384 (59)
1941	29,169 (70)
1942	62,462 (41)
1943	54,613 (54)
1944	52,811 (57)
1945	43,356 (89)
1946	48,524 (82)
1947	39,868 (102)
1948	47,095 (74)
1949	26,141 (103)
1950	36,591 (65)
1951	35,731 (45)
1952	44,946 (49)
1953	40,972 (69)
1954	41,780 (85)
1955	39,076 (97)
1956	37,316 (118)
1957	32,024 (125)
1958	41,933 (72)
1959	37,407
1960	39,614

Mileage at 12/36: 117,725
Mileage at 31/12/50: 693,724

Sheds

Perth	15/3/35
Crewe	2/11/35
Farnley Jct	4/6/38 (lan)
Huddersfield	18/6/38
Inverness	6/12/41 (loan)
Inverness	27/12/41
Carlisle Kingmoor	4/9/43
Perth	27/7/46
Carlisle Kingmoor	3/5/47
Crewe North	28/5/49 (loan)
Crewe North	25/6/49
Crewe South	1/10/49
Holyhead	5/7/52
Chester	20/9/52
Holyhead	13/6/53
Crewe North	26/9/53
Trafford Park	6/2/54
Derby	23/11/57
Neasden	10/1/59
Derby	18/6/60
Saltley	4/3/61
Oxley	10/4/65
Crewe South	11/3/67

Stored

27/2/39-3/4/39
24/4/39-4/5/39
8/5/39-22/5/39
8/3/67-15/9/67

Withdrawn w.e 16/9/67

Crewe South engine 45006 with an express freight at Watford Junction on 5 April 1952. It has a domeless boiler and riveted tender. This is another Crewe-built engine sporting an Armstrong Whitworth worksplate on frames which were ex-45224.

45007

Built as 5007 at Crewe Works 19/3/35
Renumbered 45007 w.e. 11/12/48

Improvements and modifications
11/5/38	Removal of vacuum pump
17/12/43	Steam sanding
12/2/55	Modernisation
30/1/60	Sloping throatplate boiler
30/1/60	Fitting BR ATC equipment

Repairs
24/3/36-4/4/36	LS
19/5/37-5/6/37	LS
2/5/38-11/5/38	LS
3/2/39-24/3/39	HG
20/3/40-12/4/40	LS
14/11/40-12/12/40	LS
17/3/41-31/3/41	LO
9/8/41-13/9/41	HG
30/6/42-18/8/42	HS
27/11/42-8/1/43	LS
18/11/43-17/12/43	HG
19/10/44-21/11/44	LS
4/9/45-6/10/45	LS
30/11/45-21/12/45	LO
8/2/46-4/3/46	LO
26/9/46-30/10/46	LS
29/4/47-16/5/47	LO
4/10/47-7/11/47	HS
1/11/48-11/12/48	LS
5/2/49-12/2/49	NC
17/10/49-26/11/49	G
13/11/50-25/11/50	LC
19/3/51-19/4/51	HI
12/9/51-28/9/51	LC
17/6/52-18/7/52	LI
25/5/53-27/6/53	HI
18/11/53-28/11/53	LC
20/12/54-12/2/55	G
9/8/56-1/9/56	HI
1/3/57-8/3/57	LC(EO)
29/1/58-22/2/58	LI
9/4/58-17/4/58	LC(EO)
29/7/58-6/8/58	LC(EO)
25/8/58-6/9/58	LC(EO)
20/6/59-1/7/59	LC(EO)
12/12/59-30/1/60	G
2/3/60-5/3/60	NC
26/5/61-17/6/61	LC
26/2/62-9/3/62	NC
10/8/62-31/8/62	LI
3/7/63-8/7/63	LC

Boilers
New	8824
24/3/39	8827 from 5070
13/9/41	8933 from 5175
17/12/43	8826 from 5066
26/11/49	8684 from 5170 (domed)
12/2/55	8648 from 45084 (domed)
30/1/60	12130 from 45355 (sloping throatplate)

Tenders
New	9060
15/12/36	9061
3/2/37	9060
6/7/43	9070
25/10/44	9068
14/2/55	9030
13/2/61	9827 (welded)

Mileage/(weekdays out of service)
1935	59,386 (38)
1936	61,646 (37)
1937	58,408 (39)
1938	54,464 (50)
1939	46,018 (64)
1940	62,344 (54)
1941	53,968 (71)
1942	46,794 (96)
1943	43,525 (80)
1944	50,862 (58)
1945	43,918 (86)
1946	38,744 (123)
1947	36,957 (94)
1948	43,124 (78)
1949	46,762 (82)
1950	36,950 (56)
1951	36,426 (96)
1952	35,765 (78)
1953	50,845 (64)
1954	42,583 (68)
1955	45,708 (90)
1956	44,576 (73)
1957	45,460 (55)
1958	36,833 (71)
1959	35,939
1960	34,738
1961	24,466
1962	22,572

Mileage at 12/36: 121,032
Mileage at 31/12/50: 783,870

Sheds
Inverness	19/3/35
Crewe	30/3/35 (loan)
Perth	13/7/35
Inverness	@4/4/36
Perth	27/10/45
Motherwell	15/11/47
Perth	31/1/48
Eastfield	11/2/50 (PE)
Perth	22/7/50
St Rollox	21/7/51
Carlisle Kingmoor	17/5/52
Dalry Road	20/5/57
Corkerhill	17/7/58
Hurlford	22/9/60 (loan)

Withdrawn w.e. 23/7/64

5007 pictured at Aberdeen in 1935 soon after it was built in March with domeless boiler and riveted tender and delivered to the Northern Division where it remained for its whole life. The crosshead vacuum pump and tablet exchange apparatus are clearly visible. As 45007 it was one of the last to be converted to a sloping throatplate boiler in January 1960. Photograph H.N. Shepherd, www.transporttreasury.co.uk

Perth's 45007 gained its BR number in December 1948 and, unusually for a Scottish Region engine, has 8 inch cab numbers, positioned high up on the cab side to clear the tablet catcher. The domeless boiler was replaced by a domed example in November 1949.

45008

Built as 5008 at Crewe Works 25/3/35
Renumbered 45008 w.e. 17/7/48

Improvements and modifications
22/3/38	Removal of vacuum pump
20/5/45	Steam sanding
20/11/54	Modernisation
16/1/60	Fitting BR ATC equipment
16/1/60	Sloping throatplate boiler

Repairs
14/4/36-23/4/36	LS
15/12/36-11/1/37	TRO
27/5/37-15/6/37	LS
5/3/38-22/3/38	LS
1/4/39-10/5/39	HG
24/4/40-25/4/40	LS
3/2/41-28/2/41	TRO
2/10/41-30/10/41	LS
24/4/42-30/5/42	HG
26/1/43-23/2/43	HS
14/12/43-12/1/44	LS
20/11/44-27/12/44	LS
20/6/45-2/8/45	HG
27/3/46-27/4/46	LO
28/9/46-31/10/46	HS
30/4/47-19/6/47	LS
12/6/48-16/7/48	HS
27/10/48-28/10/48	NC
10/6/49-1/7/49	LC
17/9/49-5/11/49	G
1/12/50-16/2/51	HI
26/8/52-11/11/52	HI
1/12/52-4/12/52	NC
29/1/53-27/2/53	LC
4/7/53-16/7/53	LC
20/8/53-29/8/53	LC
22/9/53-10/10/53	LC(EO)
20/9/54-20/11/54	G
6/12/54-11/12/54	NC(EO)
4/10/55-21/10/55	LC(EO)
26/4/56-2/6/56	HI
30/11/56-15/12/56	NC(EO)
28/2/57-11/3/57	NC(EO)
4/2/58-7/3/58	HI
13/4/59-24/4/59	LC(EO)
4/12/59-16/1/60	G
22/1/62-24/2/62	LI

Boilers
New	8825
10/5/39	9046 from 5116
30/5/42	9011 from 5156
2/8/45	8991 from 5036
5/11/49	8982 from 5100
20/11/54	8946 from 45086
16/1/60	12912 (sloping throatplate)

Tenders
New	9061
15/12/36	9060
3/2/37	9061
15/3/38	9255 (welded)
3/2/41	9278 (welded)
8/3/42	9062
6/12/44	9272 (welded)
2/8/45	9279 (welded)
10/10/46	9101
15/7/48	9845 (welded)
17/11/54	10683 (part-welded)
7/3/58	9029
27/5/63	9275 (welded)
14/5/64	9158

Mileage/(weekdays out of service)
1935	46,559 (21)
1936	60,980 (42)
1937	54,251 (43)
1938	55,251 (54)
1939	51,267 (48)
1940	63,848 (38)
1941	49,969 (83)
1942	63,306 (72)
1943	55,907 (51)
1944	43,714 (102)
1945	41,329 (94)
1946	38,974 (83)
1947	51,123 (69)
1948	36,971 (102)
1949	31,468 (125)
1950	46,212 (50)
1951	42,671 (74)
1952	36,852 (110)
1953	29,432 (106)
1954	34,610 (96)
1955	40,248 (69)
1956	38,978 (92)
1957	43,787 (70)
1958	38,760 (92)
1959	35,132
1960	41,761
1961	40,230
1962	32,890

Mileage at 12/36: 107,539
Mileage at 31/12/50: 794,129

Sheds
Crewe	13/4/35
Carlisle North	11/5/35
Inverness	13/7/35
Carlisle Kingmoor	22/1/44
Motherwell	4/10/47

Stored
9/5/64

Withdrawn w.e. 12/6/64

5008 at Invershin on the old Highland line north of Inverness on 10 August 1939. The front steam heating pipe has been removed for the summer. Allocated to Inverness from July 1935, 5008 had a welded tender from March 1938 but always had a domeless boiler until conversion in January 1960 to a sloping throatplate boiler.

45008 waits for the off at Carlisle on 14 April 1960. It had been fitted with a later pattern sloping throatplate boiler, with the top feed on the first barrel ring, three months earlier when it also acquired ATC. It has reverted to a riveted tender having had eleven changes up to that date. 45008 was allocated to Motherwell from October 1947 until withdrawn in June 1964. Photograph D. Forsyth, Paul Chancellor Collection.

45009

Built as 5009 at Crewe Works 22/3/35
Renumbered 45009 w.e. 15/5/48

Mileage at 12/36: 111,088
Mileage at 31/12/50: 773,348

Improvements and modifications
10/39	Removal of vacuum pump
19/3/60	Fitting BR ATC equipment

Sheds
Inverness	22/3/35
Crewe	30/3/35 (loan)
Inverness	13/7/35
Perth	31/7/37 (loan)
Inverness	14/8/37
Perth	11/5/46
Carlisle Kingmoor	17/5/47
Motherwell	4/2/50

Withdrawn w.e. 22/11/65

Repairs
4/5/36-16/5/36	LS
14/6/37-7/7/37	LS
8/10/37-27/10/38	LS
27/11/39-11/1/40	HG
17/5/40-25/5/40	LO
4/11/40-16/12/40	LS
11/8/41-3/10/41	LS
3/3/42-30/4/42	LO
19/10/42-21/11/42	LS
13/9/43-9/10/43	HG
12/9/44-5/10/44	HS
1/10/45-19/11/45	LS
22/3/46-1/5/46	LO
26/10/46-27/12/46	HG
6/4/48-11/5/48	LS
12/10/49-5/11/49	HI
3/4/50-8/4/50	LC
1/6/50-19/6/50	LC
21/7/50-15/8/50	NC
5/2/51-28/4/51	HG
22/3/52-26/4/52	HI
8/10/53-28/11/53	HI
14/6/54-15/6/54	LC(EO)
19/10/54-20/11/54	G
24/8/56-15/9/56	LI
12/11/56-29/11/56	LC(EO)
10/12/56-15/12/56	LC(EO)
18/2/57-23/2/57	LC(EO)
18/8/58-6/9/58	LI
1/2/60-19/3/60	G
8/8/60-20/8/60	HC
15/9/60-17/9/60	NC(EO)
11/5/62-29/5/62	LI
2/9/63-10/9/63	LC(EO)
25/11/63-18/1/64	LI
15/7/65-17/7/65	NC

Boilers
New	8826
11/1/40	8832 from 5019
9/10/43	8976 from 5185
29/12/46	8938 from 5172

Tenders
New	9062
8/3/42	9278 (welded)
11/5/48	9264 (welded)
?	10579 (part-welded)
30/11/51	9481 (welded)
15/11/54	9665 (welded)
25/11/63	10578 (part-welded)
30/11/64	10672 (part-welded)

Mileage/(weekdays out of service)
1935	47,395 (27)
1936	63,693 (42)
1937	60,630 (50)
1938	52,312 (56)
1939	41,835 (73)
1940	49,114 (68)
1941	51,082 (71)
1942	47,595 (93)
1943	56,751 (46)
1944	50,622 (48)
1945	44,505 (81)
1946	39,835 (113)
1947	53,644 (50)
1948	45,912 (71)
1949	28,996 (68)
1950	39,427 (84)

Below. 45009 displays a typical Scottish Region embellishment of a coloured smokebox number and 66B Motherwell shedplate. It was based north of the border from building until withdrawn in November 1965. It always carried a domeless boiler and had a welded tender from March 1942. This 1963 photograph shows the AWS protection plate under the bufferbeam and the large vacuum reservoir on the platform in front of the cab. Although it has electrification OHLW flashes its top lamp bracket has not been moved down unlike its English classmates. Photograph Paul Chancellor Collection.

45009 at Kingmoor on 11 April 1960 still showing the new paintwork applied during a General overhaul at St Rollox completed in March 1960, when it was also fitted with AWS. It always had a domeless boiler, but as with many of the Scottish engines it was paired with several tenders of each variant, and it kept the welded example shown until late-1963. 45009 spent its first decade shuffling between Inverness and Perth before moving to Kingmoor in 1947 and then on to Motherwell from February 1950 until withdrawn in November 1965. Photograph D. Forsyth, Paul Chancellor Collection.

45010

Built as 5010 at Crewe Works 27/3/35
Renumbered 45010 w.e. 22/1/49

Improvements and modifications
26/5/38   Removal of vacuum pump
20/5/45   Steam sanding

Repairs
31/3/36-10/4/36   LS
7/4/37-24/6/37   LS
17/5/38-26/5/38   LS
19/1/39-15/2/39   HS
29/5/40-5/7/40   HG
28/12/40-18/1/41   LO
27/6/41-1/8/41   LS
23/1/42-28/2/42   HG
10/11/42-12/12/42   LS
18/9/43-16/10/43   HS
26/6/44-4/8/44   LS
2/2/45-1/3/45   LO
24/7/45-8/9/45   HS
17/6/46-3/9/46   HG
24/11/47-13/12/47   LS
8/12/48-22/1/49   HI
22/8/49-21/9/49   LC
9/2/50-1/4/50   G
24/4/50-25/4/50   NC(R)
11/3/52-2/5/52   HI
9/11/53-12/12/53   LI
6/5/54-22/5/54   LC(EO)
22/12/54-29/12/54   LC(EO)
18/1/55-11/2/55   G
6/5/55-21/5/55   LC(EO)
28/5/55-15/7/55   LC(EO)
8/12/56-23/1/57   HI
23/4/58-17/5/58   HI
25/5/59-6/6/59   G
3/10/60-3/11/60   HI
6/6/52-29/6/62   LI

Boilers
New   8827
15/2/39   8936 from 5156
5/7/40   8953 from 5173
28/2/42   8827 from 5007
3/9/46   8807 from 5109
1/4/50   8915 from 5138
11/2/55   8828 from 45016
6/6/59   8994 from 45179

Tenders
New   9063
6/2/45   9218
13/3/45   9591 (welded)
7/11/51   10592 (part-welded)
10/12/53 9090
17/5/58   10783

Mileage/(weekdays out of service)
1935   49,764 (11)
1936   63,180 (26)
1937   50,219 (82)
1938   51,169 (67)
1939   54,386 (41)
1940   59,107 (52)
1941   46,863 (108)
1942   51,950 (88)
1943   56,998 (55)
1944   50,922 (70)
1945   50,313 (81)
1946   51,456 (99)
1947   56,411 (43)
1948   41,741 (67)
1949   51,543 (72)
1950   29,186 (96)
1951   36,688 (76)
1952   31,373 (74)
1953   33,622 (103)
1954   43,140 (90)
1955   28,935 (132)
1956   36,648 (77)
1957   41,974 (62)
1958   39,736 (84)
1959   43,084
1960   38,611
1961   40,480
1962   22,226

Mileage at 12/36: 112,944
Mileage at 31/12/50: 815,958

Sheds
Crewe   13/4/35
St Rollox   11/5/35
Inverness   13/7/35
Carlisle Kingmoor   8/9/45
Perth   17/8/46
Eastfield   11/2/50
Corkerhill   10/3/51
Hurlford   7/4/51

Withdrawn w.e. 29/8/63

5010 was originally allocated to St.Rollox for two months until it moved to Inverness in July 1935. As with the other class members based there it was fitted with Manson's automatic tablet exchange apparatus. The picture clearly shows the crosshead vacuum pump which was removed in May 1938. Photograph www.transporttreasury.co.uk

45011

Built as 5011 at Crewe Works 3/4/35
Renumbered 45011 w.e. 8/1/49

Improvements and modifications
10/3/38	Removal of vacuum pump
20/5/45	Steam sanding
3/1/49	Sloping throatplate boiler
?	Fitting BR ATC equipment

Repairs
4/6/36-30/6/36	LS
1/6/37-15/6/37	LO
15/2/38-10/3/38	LS
16/5/39-5/7/39	HG
8/10/40-30/12/40	LS
22/9/41-24/10/41	LS
24/3/42-3/4/42	LO
21/11/42-23/12/42	HG
27/10/43-20/11/43	HS
15/5/44-8/6/44	LO
28/6/45-1/9/45	LS
21/6/46-15/8/46	LS
27/8/47-18/10/47	LS
10/11/48-3/1/49	G
23/8/49-7/9/49	LC
20/7/50-24/8/50	HI
3/9/51-2/11/51	HI
16/2/53-6/3/53	HI
20/9/54-23/10/54	G
3/11/54-6/11/54	LC (EO)
1/11/55-18/11/55	HI ((EO)
30/11/55-1/12/55	LC (TO)
12/11/56-22/11/56	LC (EO)
2/5/57-18/5/57	HI
2/7/57-11/7/57	LC (EO)
5/8/57-10/8/57	LC (EO)
16/9/57-4/10/57	LC (EO)
27/3/58-3/4/58	LC (EO)
19/6/58-5/7/58	HI
11/7/58-16/7/58	LC (EO)
18/2/59-27/2/59	G
16/10/61-26/10/61	LC (EO)
17/11/61-18/11/61	LC (EO)
25/9/62-11/10/62	LC (EO)
10/3/64-1/5/64	LI
8/6/65-26/6/65	LC

Boilers
New		
5/7/39	8830 from 5013
23/12/42	8658 from 5025 (domed)
20/11/43	8936 from 5014
3/1/49	12876 (sloping throatplate)

Tenders
New	9064	
21/11/42	9071
22/5/44	9060
8/6/44	9071
6/7/45	9060
3/4/48	9107
28/10/49	9277 (welded)
30/10/49	9107
24/8/50	9116
19/11/55	9122
3/12/63	9064
?	9107
?	9663 (welded)
1/1/66	10584 (part-welded)

Mileage/(weekdays out of service)
Year	Mileage (weekdays out of service)
1935	43,985 (25)
1936	66,644 (30)
1937	49,740 (77)
1938	53,639 (59)
1939	43,880 (57)
1940	39,933 (101)
1941	58,610 (51)
1942	43,943 (91)
1943	57,222 (52)
1944	41,845 (51)
1945	39,777 (89)
1946	41,779 (84)
1947	44,340 (73)
1948	37,826 (94)
1949	64,599 (41)
1950	35,375 (64)

Mileage at 12/36:110,629
Mileage at 31/12/50: 763,137

Sheds
Crewe	13/4/35
Inverness	11/5/35
Perth	27/10/45
Eastfield	11/2/50
Perth	22/7/50
Carlisle Kingmoor	29/12/51
Grangemouth	16/3/52
Eastfield	24/5/54
Corkerhill	16/5/55
Carstairs	3/10/55

Withdrawn w.e. 28/12/65

Below. **45011 at Balornock on 14 March 1953 has the final boiler type with sloping throatplate and the top feed on the first barrel ring. It received this boiler in January 1949 and carried it until October 1954 when it reverted to a vertical throatplate type. Although built at Crewe it has rectangular Armstrong Whitworth worksplates evidencing the exchange of frames from 45280 which was common practice at Crewe Works from 1943 onwards. It has large 10in cab numbers but surprisingly at such a late date there are still no insignia on the tender. Photograph www.rail-online.co.uk**

45012

Built as 5012 at Crewe Works 2/4/35
Renumbered 45012 w/e 17/9/49

Improvements and modifications
18/6/38   Removal of vacuum pump
20/5/45   Steam sanding
25/12/54   Modernisation

Repairs
19/6/36-3/7/36	LS
24/4/37-22/5/37	LS
20/5/38-18/6/38	LS
23/12/38-17/1/39	HS
18/5/39-29/6/39	HO
18/9/39-20/10/39	LO
17/6/40-7/7/40	LS
4/4/41-5/5/41	LS
18/6/42-21/7/42	HS
4/9/42-10/10/42	LO
19/1/43-23/2/43	HG
17/11/43-7/1/44	HS
9/11/44-9/12/44	LS
26/2/45-31/3/45	LO
25/10/45-30/11/45	LS
3/7/46-31/8/46	HG
26/4/47-10/6/47	LO
30/12/47-18/2/48	HS
11/8/49-16/9/49	LI
25/11/50-13/1/51	G
23/2/52-27/3/52	LI
7/4/53-2/5/53	HI
2/11/54-25/12/54	G
8/12/55-29/12/55	LC (EO)
22/1/57-14/2/57	LI (EO)
27/2/57-27/2/57	LC (EO)
16/8/57-30/8/57	LC(EO)
4/11/57-13/11/57	LC(EO)
?-14/8/58	HI
7/11/58-13/11/58	LC(EO)
22/3/60-14/5/60	GEN
4/9/61-21/9/61	LC
4/12/61-13/1/62	HI

Boilers
	New 8829
29/6/39	9049 from 5119
23/2/43	9012 from 5166
31/8/46	8969 from 5162
13/1/51	8956 from 45084
25/12/54	8826 from 45085
14/5/60	8977 from 45122

Tenders
New	9067
9/11/44	9556 (welded)
31/7/46	9064
15/9/49	9818 (welded)
21/12/54	10695 (part-welded)
29/12/55	9710 (welded)
14/2/57	9229 (welded)
27/2/57	9364

Mileage/(weekdays out of service)
1935	35,521 (38)
1936	60,906 (35)
1937	56,462 (61)
1938	61,906 (55)
1939	44,759 (103)
1940	57,648 (27)
1941	53,099 (54)
1942	35,458 (97)
1943	46,951 (85)
1944	48,589 (65)
1945	45,247 (98)
1946	45,375 (76)
1947	44,924 (67)
1948	43,527 (89)
1949	40,735 (75)
1950	43,351 (65)
1951	50,798 (44)
1952	42,558 (70)
1953	47,747 (66)
1954	35,651 (80)
1955	50,343 (49)
1956	52,829 (45)
1957	40,743 (81)
1958	44,727 (77)
1959	50,051
1960	44,485

Mileage at 12/36: 96,427
Mileage at 31/12/50: 764,758

Sheds
Crewe	13/4/35
Polmadie	11/5/35
Inverness	13/7/35
Perth	13/5/39
Corkerhill	22/7/39
Inverness	2/12/39
Motherwell	8/11/47
Inverness	10/6/50
Carlisle Kingmoor	20/9/52
Carlisle Upperby	10/3/62
Carlisle Kingmoor	22/6/63
Barrow	23/4/66

Withdrawn w.e. 8/10/66

45012 waiting on the centre road at Carlisle on 20 July 1960. It always had a domeless boiler but had all three patterns of tender before returning to the riveted type in 1957. 45012 was Scottish based until 1952, moving between Kingmoor and Upperby sheds until April 1966. Photograph D. Forsyth, Paul Chancellor Collection.

45013

Built as 5013 at Crewe Works 4/4/35
Renumbered 45013 w.e. 17/4/48

Improvements and modifications
24/5/38	Removal of vacuum pump
28/11/42	Steam sanding
23/4/55	Modernisation

Repairs
20/4/36-1/5/36	LS
9/6/37-1/7/37	LS
28/4/38-24/5/38	LS
17/9/38-11/10/38	HS
17/5/39-23/6/39	HG
24/3/40-3/7/40	LS
6/2/41-28/2/41	LS
12/6/41-12/8/41	HO
15/11/41-?	
22/8/42-26/9/42	LS
27/7/43-21/8/43	HS
11/12/44-29/12/44	LS
8/6/45-13/7/45	HO
29/10/45-26/11/45	HG
22/1/46-16/2/46	LO
6/6/47-17/7/47	LS
31/3/48-17/4/48	LO
5/2/49-16/3/49	HI
2/2/50-20/3/50	HG
22/3/50-24/3/50	TO
24/5/52-14/6/52	HI
29/6/53-10/8/53	HI
12/8/53-19/8/53	NC
5/3/55-23/4/55	G
20/12/56-12/1/57	HI
10/2/58-8/3/58	HI
22/6/59-15/7/59	GEN
22/1/60-11/2/60	LC(EO)
7/9/61-4/11/61	HI
23/10/63-27/11/63	HG
29/9/65-13/11/65	HI

Boilers
New	8830
23/6/39	8937 from 5157
12/8/41	8672 from 5133 (domed)
26/11/45	8922 from 5133
25/3/50	8951 from 5043
23/4/55	8923 from 45099
11/7/59	9044 from 45086

Tenders
New	9068
9/6/43	9164
28/12/44	9720 (welded)
19/2/47	9473 (welded)
25/3/48	9511 (welded)
23/4/48	9076
22/4/55	10527 (welded)
10/2/58	9277 (welded)
8/3/58	10683 (part-welded)
27/11/63	9782 (welded)

Mileage/(weekdays out of service)
1935	48,204 (24)
1936	65,482 (27)
1937	54,331 (38)
1938	46,917 (82)
1939	46,566 (96)
1940	58,447 (55)
1941	40,174 (143)
1942	46,472 (81)
1943	54,913 (45)
1944	48,526 (59)
1945	36,615 (113)
1946	52,450 (93)
1947	38,635 (92)
1948	36,503 (75)
1949	36,336 (59)
1950	34,594 (74)
1951	41,167 (41)
1952	45,249 (64)
1953	48,789 (73)
1954	56,290 (46)
1955	45,248 (73)
1956	42,700 (68)
1957	53,072 (44)
1958	45,368 (73)
1959	49,303
1960	47,994

Mileage at 12/36: 113,686
Mileage at 31/12/50: 748,165

Sheds
Crewe	13/4/35
Inverness	11/5/35
Perth	13/5/39
Corkerhill	22/7/39
Carlisle Kingmoor	2/12/39
Inverness	11/7/40
Carlisle Kingmoor	28/8/43
Crewe North	30/4/49
Crewe South	1/10/49
Carlisle Kingmoor	6/10/51
Grangemouth	22/3/52
Carlisle Kingmoor	16/8/52
Stockport	6/1/68

Stored
7/3/66-29/9/66

Withdrawn w.e. 27/4/68

Below. **45013 at its home shed, Carlisle Kingmoor, on 30 March 1964 where it remained until transferred to Stockport in January 1968, three months before withdrawal. Always domeless apart from a brief spell in wartime, it ran with a welded tender from November 1963 having previously had four others of this type and has the final modification of lowered top lamp bracket. The Crewe-built engine has frames with an Armstrong Whitworth worksplate, ex 45325. It has the Bryson system for which a large number of engines were equipped. The only fitting was a small bracket on the cab side into which the actual exchanger was slotted.**

45014

Built as 5014 at Crewe Works 15/4/35
Renumbered 45014 w.e. 8/5/48

Improvements and modifications
22/2/41	Removal of vacuum pump
19/5/45	Steam sanding
13/6/57	Modernisation
29/8/59	Fitting BR ATC equipment

Repairs
15/5/36-29/5/36	LS
10/11/36-27/11/36	TRO
23/6/37-9/7/37	LS
21/2/38-21/3/38	LS
5/5/38-26/5/38	LO
7/9/38-6/10/38	HS
3/3/39-24/4/39	HG
17/4/40-24/5/40	HS
19/9/40-10/10/40	HO
18/4/41-25/6/41	LO
16/12/41-21/1/42	LO
26/6/42-8/8/42	LS
17/5/43-22/6/43	HG
24/4/44-19/5/44	LS
26/2/45-30/3/45	LS
17/12/45-19/1/46	LS
29/8/46-12/10/46	HG
15/9/47-4/10/47	LO
2/4/48-7/5/48	LS
29/6/48-2/7/48	NC
30/4/49-14/6/49	HG
21/12/50-19/1/51	LI
30/6/52-2/8/52	LI
3/1/53-21/1/53	HC
30/12/53-19/1/54	HG
6/8/54-18/9/54	LC (EO)
1/11/55-19/11/55	LI
10/5/57-13/6/57	HG
4/8/59-29/8/59	HI
5/4/61-17/5/61	HI
3/1/63-26/1/63	LC
7/7/64-14/8/64	INT

Boilers
New	8831
24/4/39	8944 from 5164
10/10/40	8936 from 5010
22/6/43	9049 from 5012
12/10/46	9054 from 5121
14/6/49	9010 from 5149
19/1/54	8665 from 45149 (domed)
13/6/57	8974 from 45217

Tenders
New	9069
17/5/43	9296 (welded)
25/5/43	9069
19/5/44	9068
25/10/44	9070
16/11/44	9555 (welded)
19/12/45	9083
19/5/47	9184
19/1/54	9666 (welded)
19/11/55	9548 (welded)

Mileage/(weekdays out of service)
1935	44,517 (13)
1936	54,554 (49)
1937	62,378 (29)
1938	42,703 (115)
1939	50,653 (60)
1940	46,888 (81)
1941	48,819 (97)
1942	50,611 (74)
1943	54,103 (58)
1944	54,046 (44)
1945	55,815 (70)
1946	42,159 (84)
1947	40,279 (118)
1948	40,315 (83)
1949	42,783 (83)
1950	44,268 (66)
1951	38,958 (83)
1952	39,427 (56)
1953	40,683 (60)
1954	34,427 (110)
1955	37,597 (55)
1956	36,341 (44)
1957	38,081 (78)
1958	40,063 (43)
1959	40,864
1960	43,227

Mileage at 12/36: 99,071
Mileage at 31/12/50: 774,891

Sheds
Crewe	27/4/35
Inverness	13/7/35
Carlisle Kingmoor	10/5/47 (loan)
Carlisle Kingmoor	31/5/47
Crewe North	16/4/49 (loan)
Crewe North	30/4/49
Crewe South	1/10/49
Carlisle Canal	8/10/49
Longsight	20/1/51
Carlisle Upperby	7/7/51
Northampton	15/9/51
Patricroft	9/6/56
Carnforth	15/9/56
Lancaster	2/6/62
Carnforth	23/4/66

Withdrawn w.e. 24/6/67

5014 at Balornock in original condition showing the strengthening webs and hollow axles on the coupled wheels, crosshead driven vacuum pump and plain cylinder covers. It was transferred from Crewe to Inverness in July 1935 when only three months old and remained there until 1947.

45015

Built as 5015 at Crewe Works 16/4/35
Renumbered 45015 w.e. 2/10/48

Improvements and modifications
22/2/41	Removal of vacuum pump
19/5/45	Steam sanding
3/5/57	Modernisation
3/12/59	Fitting BR ATC equipment

Repairs
27/4/36-9/5/36	LS
29/6/37-15/7/37	LS
16/6/38-8/7/38	LO
25/1/39-11/2/39	LS
11/11/39-29/12/39	HG
6/3/40-12/4/40	LO
21/10/40-15/11/40	LS
28/5/41-28/6/41	LS
5/5/42-12/6/42	HG
6/7/43-30/7/43	HS
9/6/44-1/7/44	HS
22/12/44-19/1/45	LO
25/7/45-15/9/45	LS
30/8/46-7/11/46	HG
27/8/48-1/10/48	HS
30/11/49-24/12/49	HI
3/2/50-27/2/50	LC
2/9/50-9/10/50	HG
5/9/52-11/10/52	HG
8/6/54-30/6/54	LI
5/5/56-7/6/56	HI
12/3/57-3/5/57	HG
5/11/59-3/12/59	LI
20/6/60-30/7/60	LC(EO)
30/1/61-8/3/61	HC(EO)
18/3/63-17/4/63	HG
9/7/63-10/7/63	NC

Boilers
New	8832
11/11/39	8941 from 5161
12/6/42	9057 from 5096
7/11/46	9012 from 5012
9/10/50	9051 from 5024
11/10/52	9023 from 45193
3/5/57	8980 from 45189
17/4/63	8822 (domed)

Tenders
New	9070
6/7/43	9060
22/5/44	9071
8/6/44	9060
6/12/44	9069
3/6/46	9511 (welded)
25/3/48	9473 (welded)
?	9653 (welded)

Mileage/(weekdays out of service)
1935	41,389 (7)
1936	54,162 (57)
1937	59,439 (37)
1938	54,453 (58)
1939	43,065 (86)
1940	50,296 (63)
1941	57,312 (49)
1942	57,218 (53)
1943	57,238 (51)
1944	50,993 (71)
1945	46,776 (95)
1946	32,621 (133)
1947	39,403 (91)
1948	40,608 (88)
1949	34,209 (86)
1950	29,482 (87)
1951	30,049 (40)
1952	24,114 (67)
1953	34,058 (45)
1954	35,810 (48)
1955	35,731 (67)
1956	35,414 (72)
1957	32,325 (94)
1958	36,105 (41)
1959	31,031
1960	29,539

Mileage at 12/36: 95,551
Mileage at 31/12/50: 748,664

Sheds
Crewe	27/4/35 (loan)
Inverness	13/7/35
Carlisle Kingmoor	8/9/45
Ayr	20/7/46
Carlisle Kingmoor	17/8/46
Crewe North	23/4/49 (loan)
Crewe North	28/5/49
Crewe South	1/10/49
Monument Lane	10/6/50
Bushbury	30/9/50
Edge Hill	18/11/61

Withdrawn w.e. 16/9/67

45015 with a short parcels train at Mirfield in 1967- the strange looking signal is a remnant of the speed signalling installed there by the LMS in 1932. It has a domed boiler acquired in April 1963, AWS fitted in December 1959 and a lowered top lamp bracket. The prominent rivets on the buffer beam shows it has been replaced, because the original Crewe ones had flush rivets.

45016

Built as 5016 at Crewe Works 15/5/35
Renumbered 45016 w.e. 2/10/48

Improvements and modifications

9/3/38	Removal of vacuum pump
19/5/45	Steam sanding
?	Fitting BR ATC equipment

Repairs

26/6/36-9/7/36	LS
21/1/37-8/2/37	TRO
25/5/37-4/6/37	LO
17/2/38-9/3/38	LS
1/4/38-19/4/38	LO
27/5/38-15/6/38	HO
7/9/39-1/11/39	HG
3/7/40-20/7/40	LO
13/3/41-12/4/41	LS
21/7/41-29/8/41	HO
14/10/41-13/11/41	TO
24/4/42-30/5/42	HS
19/2/43-25/3/43	LS
27/12/43-29/1/44	LS
8/1/45-17/2/45	HS
8/2/46-16/3/46	LS
24/2/47-20/3/47	LS
28/8/47-30/10/47	LO
8/2/48-26/2/48	LO
1/11/48-2/12/48	LS
28/1/50-10/3/50	G
28/7/51-23/8/51	LI
28/1/53-28/2/53	LI
16/10/53-30/10/53	LC
10/11/53-14/11/53	NC (EO)
22/1/54-6/2/54	LC
4/8/54-17/9/54	G
1/5/56-22/5/56	LC
12/4/57-10/5/57	HI
24/10/57-9/11/57	LC (EO)
3/2/58-28/2/58	HI
7/10/58-16/10/58	NC (EO)
13/4/59-28/4/59	NC (EO)
1/6/59-4/6/59	LC (EO)
29/10/59-30/12/59	G
27/6/62-17/8/62	HI
17/4/63-25/5/63	LC (EO)
24/8/64-23/9/64	LC (EO)

Boilers

New	8833
1/11/39	8939 from 5159
29/8/41	8940 from 5177
17/2/45	9014 from 5083
10/3/50	8828 from 5213

Tenders

New	9071
21/1/37	9164
16/2/37	9071
21/11/42	9064
31/7/46	9556 (welded)
?	10685 (part-welded)
13/9/54	9276 (welded)

Mileage/(weekdays out of service)

1935	37,701 (19)
1936	67,731 (28)
1937	55,480 (54)
1938	45,565 (88)
1939	47,355 (84)
1940	61,825 (30)
1941	48,878 (111)
1942	57,337 (44)
1943	54,467 (50)
1944	48,541 (65)
1945	54,366 (55)
1946	50,064 (71)
1947	37,968 (120)
1948	36,209 (85)
1949	41,955 (64)
1950	37,234 (71)

Mileage at 12/36: 105,432
Mileage at 31/12/50: 782,676

Sheds

Inverness	25/5/35
Motherwell	15/11/47
Stirling	1/10/49
Ayr	26/8/65

Withdrawn w.e. 14/7/66

45016 at Balgreen on the outskirts of Edinburgh on 4 July 1959 was always a Scottish engine and was allocated to Stirling at this date It has a domeless boiler and had recently been fitted with AWS. It is hauling the Bertram Mills circus train which has an LMS Period 3 all first leading an LNER CCT followed by open carriage trucks carrying the lorries and trailers.

45017

Built as 45017 at Crewe Works 16/5/35
Renumbered 45017 w.e 16/10/48

Improvements and modifications
22/2/41	Removal of vacuum pump
19/5/45	Steam sanding
18/6/54	Modernisation
13/3/59	Fitting BR ATC equipment

Repairs
29/5/36-12/6/36	LS
27/5/37-5/6/37	LO
29/10/37-18/11/37	LS
12/4/38-11/5/38	LO
25/4/39-2/6/39	HG
10/9/40-2/10/40	LS
18/1/41-7/3/41	LO
10/11/41-20/12/41	HG
24/8/42-3/10/42	LS
14/7/43-26/8/43	HS
30/12/44-3/2/45	HS
14/5/46-7/6/46	LS
17/3/47-16/5/47	LO
18/9/48-12/10/48	HG
21/3/50-11/4/50	HI
2/9/50-28/9/50	NC
2/12/50-30/12/50	LC
29/6/51-30/7/51	LI
7/3/52-26/4/52	HG
19/10/52-6/11/53	HI
7/5/54-18/6/54	LI
11/8/56-8/9/56	HG
31/1/59-13/3/59	HG
27/6/62-3/8/62	HI
20/2/64-7/3/64	LC (EO)
15/9/65-16/10/65	HI

Boilers
New	8834
2/6/39	8935 from 5155
20/12/41	9003 from WD
26/8/43	8919 from 5105
12/10/48	8989 from 5086
26/4/52	8918 from 45201
8/9/56	9002 from 45199
13/3/59	9041 from 45077

Tenders
New	9072
@31/12/42	9178
2/2/45	9261 (welded)
2/4/47	9530 (welded)

Mileage/(weekdays out of service)
1935	42,385 (15)
1936	60,630 (44)
1937	52,231 (60)
1938	52,068 (72)
1939	52,511 (56)
1940	62,702 (67)
1941	48,562 (100)
1942	65,330 (44)
1943	55,902 (57)
1944	52,939 (52)
1945	40,987 (87)
1946	36,927 (103)
1947	34,056 (122)
1948	36,020 (107)
1949	41,059 (36)
1950	25,975 (94)
1951	44,025 (60)
1952	46,571 (74)
1953	41,351 (77)
1954	46,625 (62)
1955	46,966 (42)
1956	39,559 (53)
1957	50,939 (30)
1958	34,770 (42)
1959	35,223
1960	34,032

Mileage at 12/36: 102,965
Mileage at 31/12/50: 760,234

Sheds
Inverness	25/5/35
Carlisle Kingmoor	28/8/43
Edge Hill	23/4/49 (loan)
Edge Hill	28/5/49
Rugby	17/2/51
Edge Hill	24/3/51
Carnforth	11/8/51
Springs Branch	2/5/59
Southport	20/7/63
Newton Heath	20/6/64
Trafford Park	14/11/64
Carnforth	21/8/65

Stored
19/4/66-18/7/66
4/9/66-26/6/67

Withdrawn w.e 10/8/68

5017 had been in service less than two weeks when recorded on 28 May 1935 at Carlisle Kingmoor. It lasted until the end of BR steam being withdrawn from Carnforth in August 1968. Crewe-built features include filled in front framing, short chimney, flush riveted bufferbeam, LMS insignia on the tender at 60in spacing, ribbed tender axlebox covers, and cab roof gutters. Photograph R.K. Blencowe.

45018

Built as 5018 at Crewe Works 17/5/35
Renumbered 45018 w.e. 18/3/50

Improvements and modifications
17/5/38   Removal of vacuum pump
20/5/45   Steam sanding

Repairs
6/4/36-17/4/36   LS
13/5/37-2/6/37   LS
9/5/38-17/5/38   LS
20/8/38-1/9/38   LO
22/2/39-2/3/39   LO
16/5/39-27/7/39   HG
27/2/40-26/4/40   LO
1/5/40-11/5/40   LO
7/12/40-2/1/41   LS
17/9/41-25/10/41   LS
18/5/42-13/7/42   HS
10/6/43-20/7/43   HG
11/4/44-12/5/44   LS
22/7/44-9/9/44   LO
14/5/45-30/6/45   LS
16/5/46-21/6/46   LS
26/3/47-9/5/47   HS
15/1/48-25/2/48   HG
4/2/49-5/3/49   LI
7/2/50-17/3/50   LI
21/10/50-25/11/50   HC
22/1/52-16/2/52   LI
6/3/52-7/3/52   NC(TO)
4/11/53-26/12/53   G
5/1/54-9/1/54   NC
25/5/55-18/6/55   HI
11/7/55-14/7/55   NC(EO)
15/5/56-16/6/56   LI(EO)
23/6/56-30/6/56   NC(EO)
17/6/57-1/8/57   HI
6/11/58-29/11/58   HG
7/1/59-21/1/59   LC(EO)
27/3/59-8/4/59   LC(EO)
11/5/59-28/5/59   LC(EO)
11/7/59-22/8/59   LC
20/10/59-21/11/59   LC
3/7/61-25/8/61   HI
7/12/61-19/1/62   LI

Boilers
New   8835
27/7/39   8828 from 5011
28/7/43   8915 from 5122
25/2/48   9004 from 5123
26/12/53 8937 from 45165
11/58   8924 from 45123

Tenders
New   9164
21/1/37   9071
16/2/37   9164
9/6/43   9068
21/6/43   9083
19/12/45 9555 (welded)
28/1/48   9829 (welded)
25/11/50 9365
17/12/53 9715 (welded)
18/6/55   9314 (welded)
16/6/56   9260 (welded)
30/6/56   10692 (part-welded)
10/7/59   9502 (welded)

Mileage/(weekdays out of service)
1935   42,154 (11)
1936   63,703 (33)
1937   57,630 (52)
1938   51,971 (71)
1939   45,652 (75)
1940   44,200 (98)
1941   55,623 (54)
1942   56,955 (60)
1943   54,041 (65)
1944   45,984 (99)
1945   45,718 (66)
1946   46,128 (58)
1947   46,425 (79)
1948   42,642 (78)
1949   42,033 (75)
1950   23,957 (125)
1951   38,223 (45)
1952   34,845 (70)
1953   22,409 (102)
1954   53,790 (45)
1955   44,147 (53)
1956   49,816 (68)
1957   45,172 (94)
1958   40,411 (86)
1959   31,604
1960   23,382

Mileage at 12/36: 105,857
Mileage at 31/12/50: 764,816

Sheds
Crewe   18/5/35
St Rollox   13/7/35
Carlisle North   27/7/35
Inverness   @17/4/36
Perth   8/9/45
Motherwell   25/10/47
Inverness   10/12/49
Aviemore   22/4/50
Carlisle Kingmoor   18/10/52
St Margarets   22/8/53
Carlisle Kingmoor   19/9/53
Warrington   21/7/62
Bolton   18/5/63
Carlisle Kingmoor   23/11/63

Withdrawn w.e. 24/12/66

45018 at Lancaster on 2 September 1962 was a Scottish loco until July 1962 when it was re-allocated to Warrington. It shows a few Scottish Region touches including an 'incorrect' top feed cover on its domeless boiler, polished rim on the smokebox numberplate and close-spaced cab numbers. It has OHL warning flashes but no AWS and has a welded tender. Photograph www.rail-online.co.uk

45019

Built as 5019 at Crewe Works 21/5/35
Renumbered 45019 w.e. 15/5/48

Improvements and modifications
23/9/39	BTH speed indicator
23/9/39	Steam sanding
?	Removal of vacuum pump
27/2/59	Fitting BR ATC equipment

Repairs
21/9/36-8/10/36	LS
21/1/37-27/1/37	LO
8/4/37-2/6/37	HO
19/2/38-28/3/38	LS
27/12/38-10/1/39	LO
1/9/39-23/9/39	HG
18/10/40-6/11/40	LO
2/8/41-23/8/41	HS
18/1/43-26/2/43	LO
3/7/43-31/7/43	HG
11/1/46-6/2/46	LS
7/12/46-3/1/47	LO
10/10/47-22/10/47	LO
5/4/48-12/5/48	HG
23/10/50-17/11/50	HI
4/10/52-8/11/52	HG
16/10/54-16/11/54	LI
18/10/56-13/11/56	HG
27/11/56-5/12/56	NC(Rect)
22/8/58-26/9/58	LI
19/2/59-27/2/59	NC(EO)
20/3/61-25/4/61	LI
2/12/63-1/2/64	HG

Boilers
New	8836
23/9/39	8637 from 5121 (domed)
31/7/43	8912 from 5143
12/5/48	8911 from 5069
8/11/52	9032 from 45134
23/11/56	8966 from 45201

Tenders
New	9165
22/2/64	9292 (welded)

Mileage/(weekdays out of service)
1935	37,710 (21)
1936	50,053 (48)
1937	41,039 (79)
1938	43,094 (68)
1939	38,031 (80)
1940	40,252 (42)
1941	34,454 (48)
1942	32,907 (20)
1943	24,772 (117)
1944	34,209 (52)
1945	26,027 (47)
1946	28,550 (60)
1947	25,691 (64)
1948	26,174 (70)
1949	27,226 (34)
1950	27,358 (54)
1951	41,794 (36)
1952	34,826 (71)
1953	49,334 (44)
1954	38,063 (62)
1955	48,814 (47)
1956	33,052 (63)
1957	51,857 (38)
1958	35,867 (55)
1959	44,368
1960	33,525

Mileage at 12/36: 87,763
Mileage at 31/12/50: 537,547

Sheds
Crewe	25/5/35
Chester	5/10/35
Patricroft	8/2/36
Shrewsbury	1/6/40 (loan)
Patricroft	29/6/40
Springs Branch	1/11/41
Preston	16/5/42
Springs Branch	24/3/45
Edge Hill	10/2/51
Carnforth	7/7/51
Springs Branch	2/5/59

Withdrawn w.e. 13/5/67

Springs Branch 45019 at Perth on 30 August 1965 with all of the later modifications in view: AWS fitted in March 1959, lowered top lamp brackets from 1963 and external steam lance pipework. It had a domeless boiler apart from one domed example carried between 1939 and 1943; the riveted tender was not recorded on the Record Card which shows a welded example fitted from February 1964.

45020

Built as 5020 at Vulcan Foundry 2/8/34
Renumbered 45020 w.e 19/6/48

Improvements and modifications
22/3/37	Sloping throatplate boiler
3/6/38	Removal of vacuum pump
15/9/39	Steam sanding
28/2/59	Fitting BR ATC equipment

Repairs
14/5/35-17/6/35	LS
29/10/35-25/11/35	LS
19/10/36-24/10/36	LO
25/1/37-25/3/37	HG
18/10/37-30/10/37	LO
19/8/38-3/6/38	HS
17/8/39-15/9/39	HG
18/3/41-29/3/41	HS
21/8/43-11/9/43	LS
15/9/45-13/10/45	HG
5/10/46-28/10/46	LS
29/5/48-15/6/48	HS
30/8/49-30/9/49	LI
13/10/50-7/12/50	HG
28/5/52-24/6/52	LI
10/4/53-18/5/53	LI
8/8/54-27/8/54	HI
30/8/54-9/9/54	NC(Rect)
17/8/55-19/9/55	HG
19/11/56-13/12/56	LI
13/12/57-31/12/57	LC(EO)
10/5/58-12/6/58	LI
5/6/58-12/6/58	NC(Rect)(EO)
2/2/59-28/2/59	LC(EO)
6/6/50-21/7/60	HG
2/9/60-24/9/60	LC(EO)
13/4/62-8/5/62	INT
16/1/63-4/2/63	LC
8/4/64-10/4/64	LC

Boilers
New	8637
22/3/37	New (sloping throatplate)
15/9/39	9396 from 5276 (sloping throatplate)
29/3/41	9742 from 5058 (sloping throatplate)
13/10/45	9526 from 5434 (sloping throatplate)
7/12/50	10140 from 5314 (sloping throatplate)
19/9/55	9402 from 45256 (sloping throatplate)
21/7/60	9363 from 45434 (sloping throatplate)

Tenders
New	9074
8/10/46	9523 (welded)
18/9/64	10258

Mileage/(weekdays out of service)
1934	26,663 (4)
1935	46,661 (125)
1936	48,584 (21)
1937	44,150 (71)
1938	48,426 (30)
1939	38,930 (18)
1940	33,532 (37)
1941	30,070 (44)
1942	35,663 (40)
1943	21,592 (46)
1944	25,034 (31)
1945	31,914 (66)
1946	35,172 (46)
1947	41,349 (35)
1948	41,651 (53)
1949	42,173 (63)
1950	38,111 (77)
1951	43,699 (41)
1952	37,773 (56)
1953	38,576 (58)
1954	32,945 (70)
1955	34,144 (72)
1956	42,189 (46)
1957	46,095 (52)
1958	43,542 (49)
1959	39,254
1960	30,098

Mileage at 12/36: 121,908
Mileage at 31/12/50: 630,475

Sheds
Perth	2/8/34
Carlisle Kingmoor	20/10/34
Edge Hill	25/11/35
Holyhead	15/2/36
Edge Hill	7/3/36
Crewe South	13/1/40
Crewe North	26/6/43
Crewe South	7/8/43
Rugby	26/4/47
Edge Hill	19/9/53
Bletchley	26/2/55
Northampton	5/11/60
Willesden	19/11/60
Bletchley	7/1/61
Willesden	8/4/61
Bletchley	9/9/61
Willesden	10/11/62
Stoke	20/4/63
Llandudno Jct	22/6/63
Stoke	29/6/63

Withdrawn w.e. 11/12/65

5021 at Perth in May 1935 where it was allocated from new until November 1935 when it moved to Edge Hill as part of the exchange of later built engines with the earliest examples with lower superheat boilers. It has the front footstep between the front frames but otherwise is still in original condition with tall chimney and raised covers on the top feed delivery pipes, and scalloped steam pipe casings. Note the bracket attached to the bottom of the front cylinder cover supporting the front ends of the drain pipes and the stiffening webs at the rear of the four spokes adjacent to the crankpins. Photograph W. Hermiston, www.transporttreasury.co.uk

45021 at Crewe North in the September 1950s still in relatively original condition with domeless boiler and riveted tender. It has even retained its original 'scalloped' steam pipe covers. It was allocated to Northampton from November 1949 and stayed there until transferred to Patricroft in 1957.

45022

Built as 5022 at Vulcan Foundry 8/8/34
Renumbered 45022 w.e. 10/7/48

Improvements and modifications
3/11/36   Sloping throatplate boiler
14/11/38   Removal of vacuum pump
12/7/47   Steam sanding
23/5/59   Fitting BR ATC equipment

Repairs

14/10/35-14/11/35	LS
23/9/36-3/11/36	HG
23/7/37-16/8/37	LO
17/10/38-14/11/38	LS
22/11/39-14/12/39	HG
27/9/40-12/10/40	LO
16/6/41-2/7/41	HS
28/9/42-17/10/42	LS
6/8/43-24/9/43	LS
29/8/44-16/9/44	HG
10/9/45-13/10/45	LS
11/2/47-20/3/47	LS
30/4/47-7/5/47	LO
8/6/48-7/7/48	HG
6/8/49-17/9/49	LI
5/10/49-14/10/49	NC(R)
14/4/50-18/5/50	LC
13/11/50-13/12/50	HI
7/1/52-2/2/52	LI
14/2/52-15/2/52	NC
13/2/53-19/3/53	G
7/6/54-25/6/54	LI
4/4/55-30/4/55	HI
1/6/55-22/6/55	NC(EO)
26/10/55-2/11/55	NC(EO)
23/1/56-28/1/56	LC(EO)
13/7/56-19/7/56	NC(EO)
5/12/56-22/12/56	LI
29/5/57-5/6/57	LC(EO)
11/11/57-21/11/57	LC
14/5/58-22/5/58	LC
4/10/58-25/10/58	G
18/5/59-23/5/59	NC(EO)
10/8/59-10/9/59	LC(EO)
30/11/59-9/1/60	HI
16/1/61-26/1/61	LC(EO)
11/9/61-10/11/61	HI

Boilers
New   8639
15/10/36   9733 New (sloping throatplate)
30/12/39   9354 from 5234 (sloping throatplate)
12/7/41   9459 from 5339 (sloping throatplate)
7/7/48   12467 New (sloping throatplate)
19/3/53   9351 from 45395 (sloping throatplate)
25/10/58   8975 from 45138

Tenders
New   9076
23/4/48   9511 (welded)
13/12/50   9261 (welded)
1/2/52   9210
5/4/55   9070
30/4/55   9669 (welded)
2/11/55   9025
22/12/56   9060
4/9/63   10716 (part-welded)

Mileage/(weekdays out of service)

Year	Mileage (weekdays out of service)
1934	22,944 (10)
1935	53,234 (95)
1936	38,691 (96)
1937	42,958 (73)
1938	46,464 (105)
1939	37,280 (62)
1940	24,820 (52)
1941	30,317 (59)
1942	34,002 (52)
1943	53,362 (65)
1944	41,623 (88)
1945	43,928 (91)
1946	45,526 (56)
1947	34,787 (104)
1948	46,573 (79)
1949	33,779 (99)
1950	37,114 (79)
1951	50,214 (49)
1952	39,172 (59)
1953	60,303 (63)
1954	54,865 (43)
1955	41,462 (65)
1956	36,734 (65)
1957	42,619 (49)
1958	31,326 (93)
1959	28,371
1960	40,793
1961	29,211
1962	37,221

Mileage at 12/36: 114,869
Mileage at 31/12/50: 667,402

Sheds
Perth	15/8/34
Edge Hill	5/10/35
Warrington	12/10/35
Edge Hill	14/12/35
Northampton	25/7/36
Rugby	24/7/37
Crewe North	1/7/39
Perth	28/11/42
Carlisle N	4/9/43
Dalry Road	14/1/50

Withdrawn 16/9/63

45022 was the first sloping throatplate boiler conversion in November 1936 and was one of several to revert to the earlier vertical throatplate boiler, in this case in October 1958. It was shedded at Dalry Road from January 1950 until withdrawn in September 1963, and in this 1954 photo it has lost its scalloped steam pipe covers and has the St Rollox style of large 10in cab numbers which are closely spaced and positioned lower than the English version level with the running plate instead of with the tender emblem. Photograph A.C. Roberts.

45022 at Strawfrank Junction just to the north of Carstairs, with domeless boiler following its reconversion from sloping throatplate in October 1958 and AWS fitted in May 1959. Other changes since its days at Perth in the earlier photograph are the small cab numbers and the return of the scalloped steam pipe casings.

45023

Built 5023 at Vulcan Foundry 8/8/34
Renumbered 45023 w.e. 23/10/48

Improvements and modifications
22/2/37	Sloping throatplate boiler
5/39	Removal of vacuum pump
?	Steam sanding
28/8/59	Fitting BR ATC equipment

Repairs
29/10/35-28/11/35	LS
18/1/37-22/2/37	HS
10/11/37-1/12/37	LO
23/12/37-11/1/38	HO
14/2/39-29/2/39	LS
31/10/39-29/11/39	HS
16/9/40-28/9/40	LO
16/4/41-14/5/41	HG
27/7/42-5/9/42	LS
23/7/43-4/9/43	HG
23/6/44-8/8/44	HS
7/7/45-8/9/45	LS
27/3/46-19/4/46	LO
13/5/46-22/6/46	LS
8/12/47-26/1/48	HG
4/10/48-18/10/48	LO
18/5/49-25/6/49	HI
5/4/50-22/4/50	LC
15/5/50-16/6/50	LC
28/7/50-8/9/50	LI
12/11/51-15/12/51	HI
17/4/52-1/5/52	LC
1/9/52-4/10/52	LC
7/1/53-7/3/53	G
18/8/53-17/9/53	LC
15/5/54-22/6/54	HI
1/7/54-14/8/54	HC(EO)
23/8/54-2/9/54	LC(EO)
24/12/54-31/12/54	LC(EO)
11/4/55-27/4/55	LC(EO)
16/7/55-20/8/55	LI
14/12/55-20/12/55	NC(EO)
3/2/56-9/2/56	NC(EO)
17/2/56-24/3/56	HC(EO)
30/7/56-25/8/56	HI
12/11/56-23/11/56	LC(EO)
11/6/57-18/6/57	LC(EO)
2/4/58-5/6/58	G
11/8/59-28/8/59	HI
17/6/60-9/7/60	LC
28/7/61-9/9/61	LI
9/4/63-3/5/63	LC(EO)

Boilers
New	8640
2/2/37	9738 new (sloping throatplate)
31/12/37	10137 new (sloping throatplate)
14/5/41	9534 from 5414 (sloping throatplate)
4/9/43	9430 from 5226 (sloping throatplate)
26/1/48	10138 from 5309 (sloping throatplate)
7/3/53	8933 from 45178
5/6/58	8945 from 45083

Tenders
New	9077
20/11/42	9257 (welded)
19/2/45	9028
13/7/45	9127
6/11/51	9713 (welded)
14/12/51	9843 (welded)
7/3/53	9317 (welded)
17/9/53	9283 (welded)
20/8/55	9721 (welded)
25/8/56	9287 (welded)
5/6/58	9554 (welded)
12/9/63	9829 (welded)

Mileage/(weekdays out of service)
1934	24,637 (6)
1935	66,021 (60)
1936	57,426 (38)
1937	43,708 (90)
1938	50,437 (61)
1939	42,683 (72)
1940	42,283 (66)
1941	53,517 (52)
1942	49,165 (90)
1943	50,027 (66)
1944	45,166 (74)
1945	38,652 (115)
1946	40,748 (109)
1947	20,184 (77)
1948	45,954 (85)
1949	41,879 (103)
1950	36,939 (116)
1951	38,513 (72)
1952	44,535 (58)
1953	41,845 (96)
1954	35,293 (121)
1955	39,014 (64)
1956	31,308 (104)
1957	35,401 (71)
1958	32,299 (92)
1959	32,685
1960	38,933
1961	27,719
1962	24,430

Mileage at 12/36: 148,084
Mileage at 31/12/50: 749,426

Sheds
Perth	20/8/34
Edge Hill	29/10/35
Carlisle W	18/4/36
Bath	30/4/38
Leeds	16/3/40
Bath	25/5/40
Perth	14/9/41
Carlisle N	28/8/43
Dalry Road	14/1/50
Polmadie	17/3/51
Dalry Road	29/9/51

Withdrawn w.e. 16/9/63

45023 was in its second spell at its home shed Dalry Road to where it returned from Polmadie in March 1951 and remained until withdrawn in September 1963. The welded tender was first attached in November 1951 and the domeless boiler in March 1953, having carried sloping throatplate boilers since 1937. It has St.Rollox 10in cab numbers and has retained its original 'scallop' shaped steam pipe covers. Photograph R.K. Blencowe.

45023 and 45030, both from Dalry Road, hauled the Royal Train from Edinburgh when the Queen travelled to Glasgow on 3 July 1958. Both engines had emerged from General overhauls a few weeks earlier but would have been further 'bulled-up' for the occasion.

45024

Built as 5024 Vulcan Foundry 5/8/34
Renumbered 45024 w.e. 15/1/49

Improvements and modifications
22/4/39	Removal of vacuum pump
26/7/45	Steam sanding
6/4/57	Modernisation
20/6/59	Fitting BR ATC equipment

Repairs
29/10/35-29/11/35	LS
15/2/37-9/3/37	HS
30/12/37-21/2/38	HG
18/3/39-27/4/39	LS
26/3/40-10/4/40	HS
19/5/41-5/6/41	HG
11/3/43-26/3/43	LS
12/4/44-28/4/44	LS
26/6/45-26/7/45	HG
2/11/45-22/11/45	LO
24/9/47-30/10/47	LS
17/12/48-13/1/49	LI
23/5/50-29/6/50	HG
15/10/51-17/11/51	HG
19/4/52-30/4/52	LC(EO)
29/4/53-6/6/53	HI
6/11/54-27/11/54	HI
1/8/55-20/8/55	LC(EO)
21/9/56-10/10/56	LC(EO)
25/2/57-6/4/57	HG
1/5/58-15/5/58	LC(EO)
2/6/59-20/6/59	HI
26/4/60-16/6/60	LC(EO)
2/9/60-27/9/60	LC(EO)
29/7/61-31/8/61	LI
19/9/63-29/10/63	HG

Boilers
New	8641
7/2/38	8663 from 5046 (domed)
5/6/41	8977 from 5122
26/7/45	9051 from 5187
29/6/50	9030 from 5028
17/11/51	8935 from 45202
6/4/57	9032 from 45019
?	8662 (domed)

Tenders
New	9078
22/11/45	9550 (welded)
?	10600 (part-welded)

Mileage
1934	31,013 (3)
1935	54,770 (93)
1936	55,782 (26)
1937	41,523 (58)
1938	49,038 (79)
1939	34,995 (51)
1940	34,601 (39)
1941	35,274 (47)
1942	34,189 (35)
1943	37,100 (38)
1944	31,163 (42)
1945	30,164 (89)
1946	36,185 (40)
1947	26,946 (82)
1948	29,871 (46)
1949	36,685 (67)
1950	29,547 (53)
1951	38,607 (71)
1952	41,088 (36)
1953	39,819 (59)
1954	33,211 (51)
1955	37,501 (45)
1956	38,869 (43)
1957	39,482 (62)
1958	42,225 (52)
1959	43,159
1960	31,898

Mileage at 12/36: 141,565
Mileage at 31/12/50: 628,846

Sheds
Perth	22/8/34
Edge Hill	29/10/35
Carlisle W	18/4/36
Patricroft	17/7/37
Longsight	25/9/37
Edge Hill	16/9/39
Crewe North	22/5/43
Crewe South	26/2/44
Willesden	30/9/44
Rugby	18/6/60
Preston	10/12/60
Crewe South	9/9/61
Springs Branch	9/12/61
Longsight	10/12/61 (loan)
Springs Branch	30/12/61

Withdrawn w.e. 13/5/67

5024 was one of the engines originally delivered to the Northern Division and replaced by 21-element superheater boiler engines in late-1935 in recognition of the more exacting duties north of the Border. It moved to the Western Division, firstly to Edge Hill and then to Carlisle, Patricroft and Longsight before returning to Edge Hill in late-1939. This picture was taken probably in 1936, about two years before it received a domed boiler.

45024 from Springs Branch runs through Lancaster on 21 August 1962 with a freight train topped with Covhops or Soda Hops. Those distinctive electrification masts are from the ex-Midland Railway route to Morecambe and Heysham which was used in 1908 for an early trial of electrification using 6600V AC at 25 Hz. After nationalisation it was again used, in 1953, as a trial site for electrification at 50Hz, the voltage remaining at 6,600 prior to the adoption of the 25kV, 50Hz system as standard for new electrification on British Railways. 45024 was equipped with AWS in June 1959 and has a domeless boiler which it appears to have carried until 1963; the welded tender was replaced by a part-welded one before it was withdrawn in May 1967. It has lost its Vulcan Foundry 'scalloped' steam pipe covers. Photograph F.A. Blencowe.

45025

Built as 5025 at Vulcan Foundry 2/8/34
Renumbered 45025 w.e. 11/2/50

Improvements and modifications

17/8/38	Removal of vacuum pump
24/10/42	Steam sanding
23/5/59	Fitting BR ATC equipment

Repairs

29/10/35-10/12/35	LS
9/10/36-1/12/36	HS
13/4/37-26/4/37	LO
14/9/37-5/10/37	HG
21/10/37-9/11/37	LO
19/7/38-17/8/38	LO
17/4/39-11/5/39	HS
13/10/39-16/12/39	HS
27/1/41-8/2/41	LS
6/9/41-24/9/41	LO
2/1/42-27/1/42	LO
24/9/42-24/10/42	HG
4/3/44-20/3/44	LS
19/4/45-23/5/45	HS
5/7/46-9/8/46	HG
10/2/47-29/3/47	HO
21/2/48-19/3/48	LS
16/1/50-9/2/50	HG
10/10/51-21/11/51	LI
4/2/52-13/3/52	LC
26/6/53-3/8/53	HI
29/12/54-29/1/55	HG
13/8/55-2/9/55	LC(EO)
6/9/56-12/10/56	LI
4/1/57-8/2/57	LC(EO)
21/1/58-30/1/58	LC(EO)
21/4/58-23/5/58	LI
13/3/59-23/5/59	LC(EO)
10/12/59-22/1/60	HG
13/7/61-29/8/61	HI
26/2/62-23/3/62	LC
29/10/63-6/12/63	HI
15/4/66-21/5/66	HI

Boilers

New	8642
21/9/37	8658 from 5041 (domed)
24/10/42	9047 from 5158
9/8/46	8947 from 5096
9/2/50	8931 from 5205
29/1/55	9027 from 45215
22/1/60	8830 from 45129
?	9013

Tenders

New	9079

Mileage/(weekdays out of service)

1934	27,121 (9)
1935	58,918 (92)
1936	51,090 (66)
1937	37,049 (93)
1938	44,961 (79)
1939	35,732 (87)
1940	32,710 (30)
1941	25,547 (65)
1942	33,766 (77)
1943	40,750 (17)
1944	26,675 (43)
1945	35,235 (59)
1946	39,075 (77)
1947	37,372 (81)
1948	38,897 (44)
1949	33,447 (48)
1950	36,955 (49)
1951	32,975 (53)
1952	37,188 (62)
1953	36,306 (52)
1954	25,537 (48)
1955	38,897 (68)
1956	34,543 (58)
1957	43,697 (48)
1958	42,844 (64)
1959	36,379
1960	44,254

Mileage at 12/36: 137,129
Mileage at 31/12/50: 635,300

Sheds

Perth	27/8/34
Edge Hill	26/10/35
Carlisle W	18/4/36
Patricroft	17/7/37
Longsight	25/9/37
Edge Hill	16/9/39
Willesden	30/9/44
Northampton	27/1/45
Crewe North	25/1/47
Bletchley	8/2/47
Crewe North	13/3/48
Longsight	5/6/48 (lan)
Crewe North	19/6/48
Crewe South	23/10/48
Willesden	18/12/48
Carlisle Upperby	12/7/58
Lancaster	22/6/63
Carnforth	23/4/66

Stored
9/2/38-12/3/38
7/8/66-29/1/68

Withdrawn w.e. 10/8/68

The oldest member of the class which has been preserved is 45025 and still with domeless boiler and its original riveted tender. It also retains two of the wheelsets with hollowed out axles but has lost the 'scalloped' steam pipes. AWS was fitted in May 1959 and it has a lowered top lamp bracket. Pictured at Rose Grove some time after reallocation to Carnforth in April 1966 from where it was withdrawn on the last day of steam, having been in store between 7/8/1966 and 29/1/1968. Photograph www.rail-online.co.uk

45025 in immaculate external condition at Carnforth on 2 August 1968, just two days before the end of BR steam when it was used on an LCGB 'Last Day of Steam' special from Carnforth to Liverpool, piloted by 45390.

45026

Built as 5026 at Vulcan Foundry 1/9/34
Renumbered 45026 w.e. 1/5/48

Improvements and modifications
24/2/37	Sloping throatplate boiler
6/4/38	Removal of vacuum pump
2/8/39	BTH speed indicator
25/9/46	Steam sanding
29/3/56	Modernisation
3/1/59	Fitting BR ATC equipment

Repairs
27/8/35-18/9/35	LS
22/6/36-4/7/36	LS
18/1/37-24/2/37	HG
19/3/38-6/4/38	HS
3/7/39-2/8/39	HS
3/7/40-18/7/40	LS
11/6/41-17/7/41	HG
22/12/41-26/1/42	LO
14/8/42-11/9/42	LS
9/12/42-8/1/43	LO
2/11/43-19/11/43	LS
28/5/45-20/6/45	HS
30/8/46-25/9/46	HG
19/12/47-27/1/48	NC
17/3/48-28/4/48	LS
20/12/49-17/1/50	LI
2/12/51-16/1/52	HG
25/3/54-14/4/54	LI
16/2/56-29/3/56	HG
27/11/58-3/1/59	HG
24/11/61-19/12/61	LI

Boilers
New	8643
11/2/37	9730 New (sloping throatplate)
17/7/41	9548 from 5428 (sloping throatplate)
25/9/46	9550 from 5333 (sloping throatplate)
16/1/52	9479 from 45310 (sloping throatplate)
29/3/56	11896 from 45276 (sloping throatplate)
3/1/59	8650 from 45078 (domed)

Tenders
New	9080

Mileage/(weekdays out of service)
1934	24,589 (6)
1935	66,715 (76)
1936	44,568 (53)
1937	46,051 (52)
1938	49,215 (43)
1939	44,056 (36)
1940	35,500 (42)
1941	28,441 (88)
1942	25,544 (97)
1943	35,008 (54)
1944	33,758 (35)
1945	31,616 (44)
1946	26,015 (64)
1947	27,634 (43)
1948	30,756 (91)
1949	32,463 (32)
1950	35,436 (42)
1951	21,642 (52)
1952	30,586 (36)
1953	31,544 (35)
1954	27,840 (60)
1955	27,840 (45)
1956	29,295 (56)
1957	29,426 (50)
1958	26,315 (57)
1959	31,580
1960	35,559

Mileage at 12/36: 135,872
Mileage at 31/12/50: 617,365

Sheds
Perth	28/8/34
Edge Hill	28/9/35
Warrington	4/8/45
Patricroft	18/12/48
Springs Branch	30/9/50
Blackpool	27/7/63
Fleetwood	9/11/63
Rose Grove	5/12/64
Patricroft	6/2/65
Newton Heath	19/6/65

Withdrawn w.e. 9/10/65

45026 on 22 August 1958 at Springs Branch where it was allocated from 1950 until 1963. The low atomiser cover is consistent with the later type of sloping throatplate boiler which it carried from February 1937 until it reverted back to the sloping throatplate type at the end of 1958. It still has its original riveted tender which now has standard cruciform pattern axlebox covers and 'scalloped' steam pipe covers. Note the damaged cylinder cover. Photograph Peter Groom.

45027

Built as 5027 at Vulcan Foundry 1/9/34
Renumbered 45027 w.e. 6/11/48

Improvements and modifications
19/12/36	Sloping throatplate boiler
22/4/39	Removal of vacuum pump
10/8/39	BTH speed indicator
2/10/43	Steam sanding
1/1/58	Modernisation
7/5/60	Fitting BR ATC equipment

Repairs
20/11/35-12/12/35	LS
7/11/36-19/12/36	HG
3/1/38-19/1/38	HS
26/7/39-10/8/39	LS
3/2/40-9/2/40	LO
18/7/40-3/8/40	HG
17/6/42-11/7/42	LS
25/6/43-17/7/43	LS
6/4/45-21/4/45	HG
22/6/46-27/7/46	LO
4/7/47-20/8/47	HS
1/10/48-4/11/48	LS
29/12/48-29/1/49	NC
3/5/50-5/6/50	HG
27/8/51-14/9/51	HI
22/4/52-15/5/52	LI
18/2/54-16/3/54	HG
26/4/56-18/5/56	LI
15/10/56-2/11/56	LC(EO)
19/11/57-1/1/58	HG
26/4/60-7/5/60	NC(EO)
3/11/60-2/12/60	LI
7/5/63-5/6/63	HG

Boilers
New	8644
3/12/36	9736 New (sloping throatplate)
3/8/40	9483 from 5363 (sloping throatplate)
21/4/45	9491 from 5301 (sloping throatplate)
5/6/50	9356 from 5059 (sloping throatplate)
16/3/54	10142 from 44800 (sloping throatplate)
1/1/58	10132 from 45299 (sloping throatplate)
5/6/63	11894 (sloping throatplate)

Tenders
New	9081

Mileage/(weekdays out of service)
1934	17,422 (25)
1935	66,788 (78)
1936	58,417 (68)
1937	37,198 (40)
1938	43,987 (44)
1939	48,306 (34)
1940	34,039 (38)
1941	31,088 (59)
1942	29,712 (62)
1943	30,476 (50)
1944	26,461 (36)
1945	28,542 (47)
1946	31,820 (85)
1947	24,682 (101)
1948	24,117 (101)
1949	24,811 (89)
1950	31,082 (57)
1951	35,538 (57)
1952	46,240 (43)
1953	35,317 (54)
1954	39,608 (46)
1955	40,549 (60)
1956	31,639 (75)
1957	36,169 (76)
1958	41,565 (30)
1959	38,257
1960	33,136

Mileage at 12/36: 142,627
Mileage at 31/12/50: 588,948

Sheds
Perth	30/8/34
Crewe	2/11/35
Shrewsbury	24/4/37 (loan)
Crewe	8/5/37
Stoke	18/6/38
Crewe North	29/3/41
Crewe South	12/7/41
Camden	28/7/45
Willesden	25/8/45
Mold Junction	23/6/62
Rugby	19/1/63
Nuneaton	7/3/64 (loan)
Aston	18/7/64
Woodford Halse	16/1/65
Bletchley	27/2/65
Carnforth	10/7/65 (loan)
Carnforth	24/7/65
Stockport	25/9/65

Withdrawn w.e. 4/5/68

Willesden allocated 45027 heads a short fitted freight on 6 July 1961 at Ashton on the WCML. It was one of the early conversions to sloping throatplate boiler, in December 1936, although it has kept its original riveted tender. It was fitted with AWS in May 1960. Photograph www.rail-online.co.uk

45028

Built as 5028 at Vulcan Foundry 8/9/34
Renumbered 45028 w.e. 21/5/49

Improvements and modifications
6/4/38	Removal of vacuum pump
?	Steam sanding
15/1/59	Fitting BR ATC equipment

Repairs
28/6/35-10/7/35	LS
25/11/35-14/12/35	LS
1/2/37-22/2/37	LS
12/4/37-26/4/37	LO
21/2/38-6/4/38	HG
5/10/39-31/10/39	LS
17/11/41-19/12/41	HG
9/6/43-23/6/43	LS
21/9/44-7/10/44	LS
11/12/45-29/12/45	HG
30/1/48-21/2/48	LS
13/4/49-18/5/49	HI
25/3/50-8/5/50	HG
10/12/51-7/1/52	LI
13/11/52-13/12/52	HC
10/6/53-29/6/53	LI
12/7/53-17/8/53	HC(EO)
9/9/54-14/10/54	HG
28/12/55-4/2/56	HC(EO)
19/3/57-13/4/57	HI
22/12/58-15/1/59	LI
29/7/60-2/9/60	HG
20/10/62-10/11/62	HI
24/11/64-16/1/65	LI
8/4/65-22/5/65	LC

Boilers
New	8645
24/3/38	8981 from 5201
19/12/41	8966 from 5073
29/12/45	9030 from 5153
8/5/50	8922 from 5013
14/10/54	8672 from 45190 (domed)
2/9/60	8983 from 45103

Tenders
New	9082
29/12/45	9523 (welded)
8/10/46	9074
10/11/62	9539 (welded)

Mileage/(weekdays out of service)
1934	18,825 (2)
1935	54,512 (62)
1936	62,141 (30)
1937	37,496 (56)
1938	35,675 (58)
1939	33,783 (58)
1940	29,461 (34)
1941	28,327 (75)
1942	34,528 (25)
1943	33,814 (37)
1944	31,078 (44)
1945	33,236 (45)
1946	38,851 (26)
1947	25,373 (81)
1948	34,313 (57)
1949	36,155 (67)
1950	41,813 (62)
1951	36,268 (42)
1952	35,690 (62)
1953	34,590 (68)
1954	36,589 (58)
1955	39,887 (56)
1956	40,526 (57)
1957	38,706 (58)
1958	35,250 (60)
1959	38,099
1960	32,541

Mileage at 12/36: 135,478
Mileage at 31/12/50: 609,381

Sheds
Perth	5/9/34
Inverness	22/9/34
Crewe South	2/11/35
Rugby	26/4/47
Crewe South	23/8/47
Warrington	1/5/48
Llandudno Jct	28/5/49
Holyhead	29/4/50
Crewe North	21/10/50
Crewe South	2/12/50
Crewe North	25/10/52
Crewe South	18/4/53
Mold Jcn	25/6/55
Holyhead	10/6/61
Mold Jct	23/9/61
Carlisle Upperby	10/11/62
Carlisle Kingmoor	22/6/63

Withdrawn w.e. 25/3/67

45028 at Polmadie some time after its transfer to Carlisle Kingmoor in June 1963 where it remained until withdrawn in March 1967. It has a domeless boiler and welded tender. It is in typical final condition with lowered top lamp bracket and AWS fitted in January 1959. Photograph N. Lester, www.transporttreasury.co.uk

45029

Built as 5029 at Vulcan Foundry 8/9/34
Renumbered 45029 w.e. 25/12/48

Improvements and modifications
19/4/38	Removal of vacuum pump
24/7/39	BTH speed indicator
27/11/48	Steam sanding
?	Fitting BR ATC equipment

Repairs
9/12/35-1/1/36	LS
21/11/36-3/12/36	LS
18/8/37-15/9/37	LS
2/10/37-26/10/37	LO
28/2/38-19/4/38	HG
29/6/39-24/7/39	HS
17/7/40-3/8/40	HG
3/1/42-28/1/42	LS
27/1/43-27/2/43	HG
5/11/43-24/11/43	HS
4/9/44-21/9/44	LS
17/2/45-3/3/45	LO
7/5/45-30/5/45	LO
10/7/45-9/8/45	LS
17/12/45-26/12/45	LO
2/5/46-25/5/46	LS
3/6/47-11/7/47	HG
18/11/48-24/12/48	HS
9/4/49-28/5/49	LC
28/10/49-2/12/49	LI
22/1/52-6/3/52	G
23/9/52-3/11/52	LC(EO)
17/11/52-12/12/52	LC(EO)
19/3/53-30/4/53	LI
8/3/54-1/4/54	LC(EO)
19/2/55-10/3/55	LI
10/5/55-19/5/55	NC(EO)
26/10/55-12/11/55	LC(EO)
13/6/56-9/8/56	G
1/4/57-6/4/57	LC(EO)
15/4/57-25/4/57	LC(EO)
21/1/58-28/1/58	NC(EO)
11/9/58-26/9/58	LI
28/3/60-30/4/60	HI
1/8/61-22/8/61	LC
14/11/62-11/12/62	G
4/12/64-5/12/64	LC
10/5/65-5/6/65	LC(EO)
19/11/65-1/12/65	NC(EO)
16/5/66-18/6/66	LC(EO)

Boilers
New	8646
4/4/38	8916 from 5136
3/8/40	9004 from 5224
27/2/43	9048 from 5148
12/7/47	8952 from 5124

Tenders
New	9083
27/11/35	9059
9/8/45	9272 (welded)
1/12/49	9025
12/11/53	9669 (welded)
9/8/56	9825 (welded)

Mileage/(weekdays out of service)
1934	21,054 (4)
1935	60,138 (58)
1936	59,470 (46)
1937	35,691 (103)
1938	43,940 (90)
1939	51,394 (50)
1940	36,516 (62)
1941	38,954 (78)
1942	46,789 (59)
1943	58,783 (62)
1944	55,687 (69)
1945	49,663 (103)
1946	51,217 (73)
1947	47,479 (106)
1948	41,218 (76)
1949	43,549 (99)
1950	50,564 (20)

Mileage at 12/36: 140,662
Mileage at 31/12/50: 792,106

Sheds
Perth	5/9/34
Inverness	22/9/34
Crewe	2/11/35
Carnforth	29/5/37
Patricroft	3/7/37
Shrewsbury	25/9/37
Bath	30/4/38
Saltley	2/3/40
Bath	25/5/40
Perth	31/10/42
Dalry Road	29/1/44
Motherwell	21/4/51

Withdrawn 19/10/66

Below. 5029 at Wick soon after entering service and still with the original Vulcan Foundry features including tall chimney, open front framing, hollow bogie axles, scalloped cut-outs on the steam pipes, prominent top feed pipes and plain tender axlebox covers. The Manson tablet exchanger and crosshead vacuum pump complete the picture. It went originally to Perth in September 1934 for two weeks before going to Inverness. It first spell in Scotland lasted only until November 1935 when it moved to Crewe although it returned there from 1942 until withdrawal after passing through several diverse sheds including Bath on the Somerset & Dorset.

45030

Built as 5030 at Vulcan Foundry 8/9/34
Renumbered 45030 w.e. 17/12/49

Improvements and modifications

20/6/38	Removal of vacuum pump
?	Steam sanding
9/1/41	BTH speed indicator
6/6/59	Fitting BR ATC equipment

Repairs

10/1/36-27/1/36	LS
7/9/36-25/9/36	LS
3/1/38-26/1/38	HG
29/4/38-20/6/38	LO
14/11/39-9/12/39	LS
7/12/40-9/1/41	HS
16/11/42-19/12/42	HG
8/1/44-27/1/44	LS
13/8/45-15/9/45	HS
11/10/47-8/11/47	HS
17/11/49-16/12/49	HG
1/6/51-23/6/51	LI
30/4/52-12/6/52	HI
7/10/52-14/11/52	HC
17/12/52-17/1/53	LC
8/5/53-23/5/53	LC(EO)
24/8/53-10/10/53	G
12/10/53-14/10/53	NC(EO)
18/11/54-11/12/54	HI
12/4/55-27/4/55	LC(EO)
1/8/56-26/8/56	LI
19/12/56-25/12/56	NC(EO)
1/5/57-9/5/57	LC
26/2/58-17/5/58	G
4/6/59-6/6/59	NC(EO)
26/10/59-28/11/59	HI
31/8/60-30/9/60	LI
20/10/60	NC(EO)
4/10/61-24/10/61	LC

Boilers

New	8647
11/1/38	8673 from 5056 (domed)
26/12/42	8640 from 5043 (domed)
15/9/45	8829 from 5193
16/12/49	8961 from 5187
10/10/53	8664 from 45194 (domed)
17/5/58	8944 from 45175

Tenders

New	9084
1/10/53	9554 (welded)
11/12/54	9679 (welded)
24/8/56	10507 (welded)
17/5/58	9192

Mileage/(weekdays out of service)

1934	18,565 (5)
1935	50,924 (57)
1936	46,270 (67)
1937	40,868 (45)
1938	35,787 (83)
1939	36,393 (64)
1940	37,549 (54)
1941	33,923 (57)
1942	31,354 (61)
1943	38,189 (29)
1944	31,020 (52)
1945	26,988 (88)
1946	24,457 (37)
1947	22,999 (57)
1948	32,619 (25)
1949	30,179 (68)
1950	40,423 (45)
1951	43,951 (47)
1952	36,112 (116)
1953	41,774 (102)
1954	46,627 (51)
1955	49,894 (49)
1956	39,789 (104)
1957	41,266 (34)
1958	29,287 (106)
1959	31,062
1960	38,501
1961	33,029
1962	24,044

Mileage at 12/36: 115,759
Mileage at 31/12/50: 578,507

Sheds

Kentish Town	8/9/34
Saltley	1/12/34
Aston	14/11/36
Rugby	6/7/40
Carlisle W	23/8/41
Crewe North	21/3/42
Mold Jcn	4/4/42
Preston	7/8/43
Springs Branch	18/11/44
Crewe South	23/10/48
Perth	6/10/51
Polmadie	24/1/53
Corkerhill	12/8/53
Polmadie	13/9/53
Corkerhill	17/7/54
Dalry Road	11/10/55

Withdrawn 29/12/62

One of the Midland Division Black 5s, 5030 was allocated originally to Kentish Town, moving three months later in December 1934 to Saltley where it remained until November 1936. It is pictured during its time at that shed; the actual location is not known, though it may well be its Birmingham 'home'. Note crosshead driven vacuum pump. 5030 was fitted with a domed boiler in January 1938.

45030 with a freight at Brinklow near Rugby in the early 1950s. It had carried a domeless boiler since 1945 and still had its original riveted tender. Allocated to Crewe South at the time of this picture, it was transferred to Perth in October 1951 and remained on the Scottish Region until withdrawn at the end of 1962. Photograph www.rail-online.co.uk

45031

Built as 5031 at Vulcan Foundry 8/9/34
Renumbered 45031 w.e. 27/11/48

Improvements and modifications

20/2/38	BTH speed indicator
19/4/39	Removal of vacuum pump
29/7/44	Steam sanding
1/4/58	Modernisation
8/8/61	Fitting BR ATC equipment

Repairs

7/2/36-20/3/36	LS
17/10/36-18/11/36	LO
3/4/37-21/4/37	LO
22/6/37-29/7/37	HG
3/3/38-29/3/38	LS
22/3/3/9-19/4/39	LS
9/11/39-6/12/39	LO
28/12/39-12/1/40	LO
17/7/40-6/8/40	HG
20/12/41-14/1/42	LS
17/12/42-22/1/43	LS
26/6/44-29/7/44	HG
3/5/45-1/6/45	LS
8/2/47-31/3/47	LS
1/11/48-25/11/48	HG
5/10/50-28/10/50	HI
6/10/52-31/10/52	HI
20/2/54-19/3/54	HG
20/12/55-14/1/56	LI
15/3/57-16/4/57	LI
27/2/58-1/4/58	HG
12/9/59-22/10/59	HI
24/7/61-8/8/61	NC(EO)
26/9/62-27/10/62	LI
17/1/63-7/2/63	LC
24/8/64-7/1/65	LC

Boilers

New	8648
19/7/37	8676 from 5059 (domed)
6/8/40	8682 from 5021 (domed)
29/7/44	8925 from 5051
25/11/48	8955 from 5197
19/3/54	8912 from 45196
1/4/58	8679 from 45068 (domed)
?	8828

Tenders

New	9085
5/9/64	9520 (welded)

Mileage/(weekdays out of service)

1934	16,241 (24)
1935	56,275 (29)
1936	41,046 (100)
1937	50,812 (81)
1938	48,164 (103)
1939	39,660 (105)
1940	33,690 (66)
1941	35,194 (56)
1942	39,309 (42)
1943	44,019 (45)
1944	23,289 (65)
1945	29,905 (59)
1946	27,705 (76)
1947	28,953 (66)
1948	25,769 (83)
1949	36,684 (40)
1950	22,176 (55)
1951	25,683 (35)
1952	24,458 (60)
1953	38,328 (40)
1954	36,470 (56)
1955	25,953 (122)
1956	42,807 (49)
1957	39,385 (57)
1958	42,019 (67)
1959	38,436
1960	39,486

Mileage at 12/36: 113,562
Mileage at 31/12/50: 598,891

Sheds

Kentish Town	8/9/34
Nottingham	27/10/34
Saltley	1/12/34
Derby	11/9/37
Belle Vue	6/11/43
Newton Heath	16/4/56
Mold Jcn	23/1/60 (loan)
Mold Jcn	6/2/60
Rugby	19/1/63
Chester (M)	22/6/63
Speke Junction	18/3/67 (loan)
Speke Junction	1/4/67

Withdrawn w.e. 1/7/67

5031 brand new at Crewe in September 1934 with hardly a mark on its buffers and yet to receive its Kentish Town shedplate. The open front footplate of the first 25 Vulcan Foundry engines is apparent.

45031 is reflected in the oily water in the turntable pit at Shrewsbury in 1966 with a breakdown train made up of GWR and LNWR vehicles in the background indicating the joint nature of the shed. The front footplate has a rectangular mark left by Armstrong Whitworth builders plates, indicating that it exchanged frames during an overhaul at Crewe. It has reverted to a domeless boiler in its final years and this photograph provides a good view of the cladding joint on the top. It was a late recipient of AWS in August 1961 and the welded tender replaced its original tender in September 1964. Photograph D.J Clarke, www.rail-online.co.uk

45032

Built as 5032 at Vulcan Foundry 15/9/34
Renumbered 45032 w.e 20/8/49

Improvements and modifications

18/7/38	Removal of vacuum pump
10/11/45	Steam sanding
1/5/59	Fitting BR ATC equipment

Repairs

6/12/35-30/12/35	LS
4/12/36-19/12/36	LO
24/3/37-3/5/37	HS
7/6/38-18/7/38	HG
27/11/39-13/12/39	LS
19/8/41-4/9/41	HG
19/12/42-7/1/43	LS
7/2/44-19/2/44	LS
17/10/44-28/10/44	LO
23/10/45-10/11/45	HG
19/7/46-10/8/46	LO
7/3/47-22/4/47	LO
23/2/48-24/3/48	HS
22/7/49-17/8/49	HG
3/5/51-28/5/51	LI
1/12/51-18/1/52	LC
21/2/53-16/3/53	LI
15/1/54-15/2/54	HG
7/4/56-8/5/56	HI
2/11/57-30/11/57	LI
30/3/59-1/5/59	HG
11/5/59-20/5/59	LC
5/1/61-3/2/61	HI

Boilers

New	8649
21/6/38	8967 from 5187
4/9/41	8638 from 5094 (domed)
10/11/45	8640 from 5030 (domed)
17/8/49	8930 from 5046
15/2/54	9015 from 45146
1/5/59	8925 from 45038

Tenders

New	9086

Mileage/(weekdays out of service)

1934	18,522 (25)
1935	49,993 (68)
1936	53,524 (53)
1937	44,814 (73)
1938	42,264 (79)
1939	42,350 (63)
1940	40,800 (40)
1941	37,586 (57)
1942	35,952 (37)
1943	35,509 (32)
1944	35,004 (50)
1945	29,493 (36)
1946	28,270 (63)
1947	24,957 (87)
1948	35,543 (62)
1949	37,420 (74)
1950	38,110 (32)
1951	34,796 (76)
1952	37,165 (56)
1953	29,951 (64)
1954	34,463 (73)
1955	43,951 (28)
1956	37,118 (56)
1957	35,369 (63)
1958	44,697 (32)
1959	34,866
1960	38,974

Mileage at 12/36: 122,039
Mileage at 31/12/50: 630,111

Sheds

Crewe	15/9/34
Northampton	6/10/34
Crewe	19/1/35
Gloucester	16/2/35
Shrewsbury	29/11/36
Willesden	20/8/38
Edge Hill	23/8/41
Preston	20/3/43 (loan)
Edge Hill	10/4/43
Warrington	4/8/45
Edge Hill	14/6/58
Speke Jcn	3/4/61

Withdrawn w.e. 8/2/64

Below. 45032 was shedded at Edge Hill from June 1958 until April 1961 and was fitted with AWS in May 1959. It has a domeless boiler and its original riveted tender. Photograph R.K. Blencowe.

45033

Built as 5033 at Vulcan Foundry 15/9/34
Renumbered 45033 w.e. 27/11/48

Improvements and modifications
12/4/38	Removal of vacuum pump
24/7/39	BTH speed indicator
16/5/42	Steam sanding
24/3/56	Modification
29/4/60	Fitting BR ATC equipment

Repairs
11/1/36-24/1/36	LO
5/3/36-26/3/36	LS
23/2/37-15/3/37	HS
19/3/38-12/4/38	HG
7/6/39-24/7/39	HS
22/8/40-12/9/40	LS
10/4/42-16/5/42	HG
4/11/43-20/11/43	LS
2/2/45-3/3/45	LS
20/2/46-29/3/46	LO
7/1/47-25/1/47	HG
2/11/48-24/11/48	HS
13/12/48-21/12/48	NC
31/5/50-19/6/50	HI
4/1/51-5/2/51	LC
21/8/52-27/9/52	HG
27/3/54-23/4/54	LI
3/2/55-26/2/55	LC(EO)
14/2/56-24/3/56	HG
4/3/57-26/3/57	LC(EO)
18/3/58-12/4/58	LI
21/4/60-29/4/60	NC(EO)
2/8/60-23/9/60	HG
25/10/61-17/11/61	LC
27/12/62-9/2/63	HI
3/6/63-14/6/63	NC

Boilers
New	8650
28/3/38	8972 from 5192
16/5/42	8670 from 5113 (domed)
25/1/47	8644 from 5113 (domed)
27/9/52	8968 from 45091
24/3/56	9028 from 45195
23/9/60	9056 from 45093

Tenders
New	9087
23/12/62	10528 (welded)
27/8/65	9252 (welded)
26/10/65	10528 (welded)

Mileage/(weekdays out of service)
1934	17,876 (13)
1935	46,873 (157)
1936	43,510 (107)
1937	44,950 (50)
1938	46,480 (38)
1939	42,408 (52)
1940	34,636 (51)
1941	23,085 (37)
1942	28,747 (60)
1943	29,433 (47)
1944	27,025 (24)
1945	30,787 (71)
1946	33,443 (71)
1947	36,560 (60)
1948	36,697 (62)
1949	40,959 (60)
1950	41,514 (57)
1951	41,491 (53)
1952	33,674 (68)
1953	46,957 (45)
1954	44,004 (59)
1955	38,519 (55)
1956	47,128 (52)
1957	43,629 (51)
1958	46,429 (42)
1959	41,996
1960	33,832

Mileage at 12/36: 108,259
Mileage at 31/12/50: 614,983

Sheds
Crewe	15/9/34
Northampton	6/10/34
Crewe	19/1/35
Derby	16/2/35
Saltley	7/11/36
Edge Hill	29/11/36
Holyhead	25/5/40 (loan)
Edge Hill	29/6/40
Crewe South	13/6/42
Rugby	26/4/47
Crewe North	24/3/51
Carlisle Upperby	18/8/51
Crewe North	7/2/53
Carlisle Upperby	13/10/56 (loan)
Crewe North	20/10/56
Longsight	6/7/57
Crewe North	21/9/57
Crewe South	6/1/62
Crewe North	2/3/63
Crewe South	14/9/63

Withdrawn w.e. 24/12/66

Below. **45033 at Lichfield in the early 1960s with a domeless boiler after a spell domed up to September 1952. The AWS was fitted in April 1960, it was allocated to Crewe North until January 1962 and had a riveted tender until paired with a welded type in December 1962. Photograph www.rail-online.co.uk**

45034

Built as 5034 at Vulcan Foundry 15/9/34
Renumbered 45034 w.e. 24/4/48

Improvements and modifications
19/5/38	Removal of vacuum pump
24/4/48	Steam sanding
16/2/59	Fitting BR ATC equipment

Repairs
14/4/36-6/5/36	LS
29/3/37-14/4/37	HS
25/2/38-19/5/38	HG
12/1/40-9/2/40	LS
28/8/41-18/9/41	HG
20/5/42-20/6/42	LO
26/11/43-16/12/43	LS
6/10/44-11/11/44	HS
13/7/46-30/7/46	LS
22/3/48-24/4/48	HG
14/7/49-5/8/49	HI
11/9/50-29/9/50	HI
17/3/52-9/4/52	HI
18/6/53-25/7/53	HG
24/10/54-17/11/54	LI
29/8/56-12/10/56	LI
14/3/57-12/4/57	LC(EO)
17/9/57-12/10/57	LC(EO)
25/10/58-27/11/58	HG
10/2/59-16/2/59	NC(EO)
29/6/60-10/8/60	HI
7/12/61-11/1/62	INT
16/1/62-25/1/62	NC
25/4/62-10/5/62	LC
13/4/65-8/5/65	LI

Boilers
New	8651
29/4/38	8978 from 5198
18/9/41	8663 from 5024 (domed)
11/11/44	8986 from 5131
24/4/48	8650 from 5114 (domed)
25/7/53	8971 from 45184
27/11/58	8678 from 45219 (domed)

Tenders
New	9088
23/6/54	9106

Mileage/(weekdays out of service)
Year	Mileage (out of service)
1934	16,288 (9)
1935	51,924 (86)
1936	55,647 (74)
1937	39,645 (42)
1938	32,938 (88)
1939	38,715 (47)
1940	28,159 (89)
1941	28,354 (46)
1942	31,972 (51)
1943	30,651 (49)
1944	30,475 (52)
1945	33,870 (34)
1946	29,390 (61)
1947	29,863 (85)
1948	36,277 (64)
1949	33,552 (63)
1950	39,978 (57)
1951	41,707 (48)
1952	36,498 (40)
1953	34,163 (69)
1954	40,220 (42)
1955	41,283 (54)
1956	28,235 (86)
1957	36,865 (94)
1958	26,892 (108)
1959	42,036
1960	28,740

Mileage at 12/36: 123,859
Mileage at 31/12/50: 587,698

Sheds
Crewe	15/9/34
Northampton	6/10/34
Crewe	19/1/35
Sheffield	10/2/35
Aston	14/11/36
Bescot	19/4/41
Crewe South	10/5/41
Crewe North	26/6/43
Crewe South	7/8/43
Rugby	26/4/47
Crewe South	27/2/54
Aston	25/6/55
Monument Lane	3/11/56
Longsight	29/10/60
Crewe North	26/11/60
Lancaster	10/6/61
Speke Jcn	9/9/61
Lostock Hall	20/6/64
Speke Jcn	11/7/64

Withdrawn w.e. 3/2/68

Below. 45034 waits for the parcels to be loaded at Harrow & Wealdstone in the early 1950s. It was at 2A Rugby until February 1954, kept the domed boiler until July 1953 and still had its original riveted tender. Photograph R.K. Blencowe.

45035

Built as 5035 at Vulcan Foundry 22/9/34
Renumbered 45035 w.e. 31/7/48

Improvements and modifications

23/5/38	Removal of vacuum pump
21/8/46	Steam sanding
22/5/59	Fitting BR ATC equipment

Repairs

16/10/35-6/11/35	LS
14/5/36-4/7/36	LS
18/3/37-9/4/37	HS
14/3/38-23/5/38	HG
27/11/39-19/12/39	HS
22/8/41-4/10/41	HG
4/2/43-2/3/43	LS
29/10/43-3/12/43	LO
13/3/45-13/4/45	LS
28/7/46-21/8/46	HG
19/6/48-30/7/48	LS
21/1/49-21/2/49	LC
5/3/50-27/3/50	HI
25/8/50-19/9/50	LC
13/7/51-22/8/51	HG
27/10/51-12/11/51	LC(EO)
11/6/53-3/7/53	LI
3/12/54-24/12/54	HG
2/8/56-7/9/56	HI
2/8/57-24/8/57	LI
20/4/59-22/5/59	HI
1/7/60-25/8/60	HG

Boilers

New	8652
10/5/38	8962 from 5182
4/10/41	8980 from 5094
21/8/46	8942 from 5160
22/8/51	9038 from 45206
24/12/54	9029 from 45198
25/8/60	9040 from 45211

Tenders

New	9089

Mileage/(weekdays out of service)

1934	13,943 (6)
1935	62,966 (112)
1936	52,340 (107)
1937	40,291 (89)
1938	41,185 (99)
1939	38,185 (73)
1940	37,382 (32)
1941	23,627 (96)
1942	33,375 (31)
1943	35,007 (76)
1944	24,205 (32)
1945	28,006 (58)
1946	26,274 (56)
1947	30,477 (58)
1948	27,686 (78)
1949	21,904 (110)
1950	32,987 (79)
1951	30,675 (80)
1952	47,313 (36)
1953	39,552 (59)
1954	35,485 (58)
1955	47,558 (20)
1956	34,740 (56)
1957	40,009 (50)
1958	34,891 (43)
1959	35,467
1960	32,164

Mileage at 12/36: 129,249
Mileage at 31/12/50: 569,840

Sheds

Crewe	22/9/34
Willesden	6/10/34
Bushbury	12/1/35
Trafford Park	10/2/35
Llandudno Jct	14/11/36
Bangor	7/1/39
Llandudno Jct	18/3/39
Crewe	13/4/40
Rugby	5/6/43
Willesden	2/10/43
Edge Hill	9/10/43
Warrington	4/8/45
Blackpool	22/6/63
Fleetwood	14/9/63

Withdrawn w.e. 21/11/64

Below. 45035 from Warrington Dallam at Springs Branch in the early 1960s. It was always domeless and kept its original riveted tender until withdrawn in November 1964. AWS was fitted in June 1959. Photograph www.rail-online.co.uk

45036

Built as 5036 at Vulcan Foundry 22/9/34
Renumbered 45036 w.e. 15/1/49

Improvements and modifications
28/3/38	Removal of vacuum pump
12/6/38	BTH speed indicator
?	Steam sanding
3/6/59	Fitting BR ATC equipment

Repairs
10/6/35-16/7/35	LO
9/12/35-16/1/36	LS
27/10/36-20/11/36	LS
15/9/37-5/10/37	LS
18/2/38-28/3/38	HG
13/10/38-2/11/38	LO
11/3/39-28/4/39	LS
10/7/40-9/8/40	HG
16/5/41-11/6/41	HS
14/8/42-19/9/42	HS
15/5/43-16/6/43	LS
28/8/43-7/12/43	LO
3/7/44-27/7/44	LS
8/3/45-14/4/45	HG
3/4/46-27/4/46	LS
22/1/47-6/3/47	LS
8/11/47-26/12/47	HG
29/11/48-12/1/49	HI
14/12/49-25/1/50	LI
16/5/51-23/6/51	LI
17/11/52-27/12/52	G
16/2/53-20/2/53	NC(EO)
27/4/53-2/5/53	LC
18/1/54-5/2/54	HI
25/10/54-6/11/54	LC(EO)
2/5/55-8/6/55	LI
1/2/56-8/2/56	LC(EO)
10/7/56-11/7/56	NC(EO)
27/12/56-19/1/57	G
15/1/58-18/1/58	LC(EO)
5/5/58-11/6/58	LI
1/6/59-3/6/59	NC
28/4/60-20/5/60	HI
14/9/60-15/9/60	NC(EO)
20/3/61-22/3/61	LC(EO)
28/4/61-2/5/61	LC(EO)

Boilers
New	8653
14/3/38	8917 from 5137
1/8/40	8991 from 5211
4/4/45	8656 from 5090 (domed)
6/12/47	8999 from 5136
27/12/52	8656 from 45127 (domed)
19/1/57	9009

Tenders
New	9090
15/12/52	9724 (welded)
19/1/57	10506 (welded)

Mileage/(weekdays out of service)
1934	14,198 (15)
1935	51,757 (130)
1936	56,735 (129)
1937	42,874 (166)
1938	48,474 (117)
1939	43,848 (100)
1940	45,249 (78)
1941	52,262 (55)
1942	58,312 (62)
1943	39,138 (137)
1944	47,836 (61)
1945	49,176 (92)
1946	55,236 (64)
1947	45,356 (113)
1948	53,019 (47)
1949	47,108 (70)
1950	49,013 (46)

Mileage at 12/36: 122,690
Mileage at 31/12/50: 799,592

Sheds
Crewe	22/9/34
Willesden	6/10/34
Bushbury	12/1/35
Kentish Town	10/2/35
Trafford Park	24/8/35
Leeds	7/10/39
Nottingham	3/7/41
Perth	14/9/41
St Margarets	28/1/50
Dalry Road	4/11/50
Polmadie	9/12/50
St Margarets	27/1/51 (PE)
Polmadie	24/3/51 (PE)
St Margarets	22/9/51
Polmadie	26/4/52
Corkerhill	12/9/53
Polmadie	19/9/53
Corkerhill	17/7/54
Dalry Road	11/10/55

Stored
8/1/62

Withdrawn w.e. 29/12/62

Below. Allocated to Perth since 1941, 45036 is in unlined black at Balornock on 29 January 1950. It has large, possibly hand painted, cab numbers in the high position following its renumbering in January 1949. When pictured it was domeless between two periods when it had domed boilers and its original riveted tender has no BR insignia.

45037

Built as 5037 at Vulcan Foundry 22/9/34
Renumbered 45037 w.e. 8/5/48

Improvements and modifications
21/6/39	Removal of vacuum pump
11/1/41	BTH speed indicator
1/8/42	Steam sanding
8/8/59	Fitting BR ATC equipment

Repairs
12/3/36-24/4/36	LS
1/3/37-19/3/37	HS
6/10/37-15/11/37	HG
5/5/39-21/6/39	HS
26/12/40-11/1/41	LS
26/6/42-1/8/42	HG
5/11/42-28/11/42	LO
30/3/44-15/4/44	LS
6/4/45-2/5/45	HS
9/9/46-28/10/46	HG
6/4/48-4/5/48	LS
25/1/50-8/2/50	HI
10/2/52-4/4/52	HG
23/11/53-22/12/53	LI
3/10/55-26/10/55	LI
15/8/57-20/9/57	HG
26/6/59-8/8/59	HI
1/6/60-29/7/60	HI
3/1/61-4/2/61	LC(EO)
11/5/62-5/6/62	HI
9/1/64-1/2/64	G
10/2/64-15/2/64	NC
21/12/64-31/12/64	Rect

Boilers
New	8654
9/10/37	8648 from 5031 (domed)
1/8/42	9010 from 5118
28/10/46	8959 from 5151
4/4/52	9035 from 45048
20/9/57	8821 from 45111 (domed)

Tenders
New	9091
21/6/39	9639 (welded)

Mileage/(weekdays out of service)
1934	14,874 (6)
1935	58,485 (70)
1936	61,824 (86)
1937	41,594 (74)
1938	50,726 (30)
1939	30,278 (63)
1940	35,254 (41)
1941	32,228 (74)
1942	24,082 (89)
1943	40,267 (32)
1944	30,266 (54)
1945	27,479 (43)
1946	28,506 (98)
1947	34,109 (29)
1948	42,425 (43)
1949	33,833 (33)
1950	32,436 (31)
1951	27,485 (47)
1952	38,025 (83)
1953	36,567 (55)
1954	37,440 (35)
1955	38,260 (65)
1956	43,145 (35)
1957	41,207 (55)
1958	37,834 (57)
1959	39,570
1960	33,646

Mileage at 12/36: 135,183
Mileage at 31/12/50: 618,666

Sheds
Crewe	22/9/34
Willesden	6/10/34
Bushbury	12/1/35
Leeds	10/2/35
Patricroft	20/12/36
Carnforth	5/11/38
Preston	3/12/38
Patricroft	17/12/38
Shrewsbury	1/6/40 (loan)
Patricroft	31/8/40
Crewe South	27/3/43
Crewe North	1/5/43
Crewe South	4/3/44
Patricroft	21/6/47
Carlisle Upperby	11/12/54
Carnforth	21/1/56
Longsight	7/11/59
Crewe North	9/1/60
Bletchley	6/5/61
Willesden	10/11/62
Stoke	20/4/63

Withdrawn w.e. 27/11/65

Below. 45037 on 7 July 1963 under the new overhead masts awaiting their catenary at Ashton with a down fitted freight. It carried a domed from September 1957, AWS from August 1959 and had been shedded at Bletchley since May 1961. It acquired the welded tender in June 1939. Compare with the view on page 77. Photograph www.rail-online.co.uk

45038

Built as 5038 at Vulcan Foundry 29/9/34
Renumbered 45038 w.e. 11/12/48

Improvements and modifications

?	Removal of vacuum pump
9/9/44	Steam sanding
20/2/58	Modification
23/10/59	Fitting BR ATC equipment

Repairs

17/12/35-17/1/36	LS
11/1/37-26/1/37	LS
14/4/37-29/4/37	LO
20/8/37-22/9/37	HG
22/11/38-28/1/39	LS
23/11/39-30/12/39	HS
6/5/41-23/5/41	LS
1/12/42-26/12/42	LS
23/8/44-8/9/44	HG
9/7/46-10/8/46	LS
6/11/48-6/12/48	HG
1/9/50-6/10/50	LI
31/3/52-25/4/52	HI
1/5/52-10/5/52	NC(Rect)
18/12/53-26/1/54	HG
1/4/55-12/5/55	LI
3/12/56-1/1/57	LI
7/2/57-6/3/57	LC(EO)
28/1/58-20/2/58	HG
5/10/59-23/10/59	NC(EO)
1/2/60-3/3/60	HI
29/9/61-7/11/61	HI
2/6/64-24/10/64	HI

Boilers

New	8655
8/9/37	8668 from 5051(domed)
23/11/39	9034 from 5104
8/9/44	8983 from 5134
6/12/48	9039 from 5208
26/1/54	8925 from 45070
20/2/58	9017 from 45187

Tenders

New	9092
26/3/52	9082
7/3/63	10484 (welded)

Mileage/(weekdays out of service)

1934	17,392 (8)
1935	50,433 (70)
1936	43,562 (87)
1937	32,489 (89)
1938	38,926 (81)
1939	33,385 (85)
1940	34,295 (43)
1941	31,331 (44)
1942	28,992 (67)
1943	30,090 (40)
1944	28,186 (58)
1945	24,751 (58)
1946	27,320 (54)
1947	31,235 (44)
1948	24,502 (81)
1949	38,323 (41)
1950	34,558 (58)
1951	39,190 (36)
1952	36,068 (54)
1953	37,815 (59)
1954	40,966 (48)
1955	38,232 (69)
1956	37,024 (62)
1957	39,932 (60)
1958	37,638 (56)
1959	36,434
1960	38,829

Mileage at 12/36: 111,387
Mleage at 31/12/50: 549,770

Sheds

Crewe	29/9/34
Saltley	10/2/35
Trafford Park	30/5/36
Saltley	6/6/36
Aston	29/11/36
Bushbury	27/2/37
Walsall	10/7/37
Monument Lane	19/2/38
Llandudno Jct	6/8/38
Aston	24/9/38
Crewe North	10/5/41
Carnforth	8/1/49 (loan)
Crewe South	12/2/49
Carlisle Upperby	25/6/55
Aston	9/7/55
Monument Lane	20/9/58
Llandudno Jct	2/7/60
Aston	8/10/60
Tyseley	2/10/65
Stoke	29/1/66
Carlisle Kingmoor	24/6/67
Stockport	6/1/68

Stored

17/4/67-1/6/67
12/6/67-28/7/67

Withdrawn w.e. 17/2/68

Below. 45038 pictured in the winter snow in the Birmingham suburbs. It was allocated to Aston from October 1960 until September 1965 when it briefly went to Tyseley before moving to Stoke in January 1966. It had a domeless boiler from November 1939 until withdrawal. 45038 is in its final condition with lowered top lamp bracket, external steam lance pipework and AWS. Photograph www.rail-online.co.uk

45039

Built as 5039 at Vulcan Foundry 6/10/34
Renumbered 45039 w.e. 20/8/49

Improvements and modifications
19/5/38	Removal of vacuum pump
26/9/42	Steam sanding
13/10/56	Modification
10/1/59	Fitting BR ATC equipment

Repairs
31/1/35-25/2/35	LO
28/1/36-10/2/36	LS
6/10/36-17/10/36	LO
28/12/36-11/1/37	LS
16/4/37-30/4/37	LO
24/2/38-19/5/38	HG
2/1/40-30/1/40	LS
13/2/41-1/3/41	LS
24/8/42-26/9/42	HG
27/3/44-13/4/44	LS
13/8/45-19/9/45	LS
7/6/46-29/6/46	HG
17/1/48-10/2/48	LS
30/7/49-18/8/49	HI
11/1/51-29/1/51	HG
7/6/52-23/7/52	LI
27/12/52-16/1/53	LC(EO)
18/6/53-7/7/53	LI
1/8/53-26/8/53	LC(EO)
18/2/54-13/3/54	LC(EO)
29/4/55-28/5/55	HI
15/6/55-27/6/55	NC(Rect)EO
26/3/56-28/4/56	LC(EO)
11/9/56-13/10/56	HG
5/12/58-10/1/59	LI
27/2/61-15/4/61	HG
7/9/64-31/10/64	HI
11/1/65-29/1/65	Rect
27/4/66-15/5/66	HC

Boilers
New	8656
17/4/38	8645 from 5028 (domed)
26/9/42	8646 from 5095 (domed)
29/6/46	8647 from 5135 (domed)
29/1/51	8639 from 5052 (domed)
13/10/56	8948 from 45091
15/4/61	8836 from 45224

Tenders
New	9093
23/7/52	9419

Mileage/(weekdays out of service)
Year	Mileage (out of service)
1934	17,559 (11)
1935	48,690 (56)
1936	45,471 (47)
1937	33,292 (84)
1938	31,931 (94)
1939	36,576 (50)
1940	32,451 (45)
1941	34,740 (40)
1942	25,737 (78)
1943	41,983 (24)
1944	34,607 (37)
1945	31,476 (62)
1946	36,826 (44)
1947	45,388 (35)
1948	40,028 (45)
1949	40,664 (49)
1950	45,284 (41)
1951	44,511 (37)
1952	33,141 (65)
1953	38,416 (77)
1954	43,672 (44)
1955	38,820 (76)
1956	37,641 (74)
1957	49,454 (30)
1958	42,109 (50)
1959	47,832
1960	40,832

Mileage at 12/36: 111,720
Mleage at 31/12/50: 622,703

Sheds
Crewe	29/9/34
Carlisle	12/1/35
Saltley	16/2/35
Trafford Park	30/5/36
Saltley	6/6/36
Aston	14/11/36
Monument Lane	27/3/37
Aston	1/5/37
Edge Hill	17/8/40
Springs Branch	9/3/46 (loan)
Carnforth	4/5/46
Edge Hill	11/8/51

Withdrawn w.e. 26/8/67

A cavalcade of Black Fives led by Carnforth's 45039 on 3 February 1951 at Dalry Road after completion of a Heavy General repair at St.Rollox when it had its last domed boiler fitted. It appears to be on a running-in trial given the chalk marks and lack of the final coat of paint on the smokebox door.

45040

Built as 5040 at Vulcan Foundry 6/10/34
Renumbered 45040 w.e. 6/11/48

Improvements and modifications

25/11/36	Sloping throatplate boiler
30/3/39	Removal of vacuum pump
3/2/45	Steam sanding
22/10/54	Modernisation
24/3/63	Fitting Smith-Stone speedometer

Repairs

6/11/35-10/12/35	LS
30/6/36-13/7/36	LO
8/10/36-4/12/36	HG
15/12/36-18/1/37	HO
5/6/37-1/7/37	LO
28/10/37-15/11/37	LS
4/4/38-4/5/38	LO
3/8/38-23/8/38	LO
2/3/39-30/3/39	LS
29/5/39-12/7/39	LO
8/11/39-16/11/39	LO
5/3/40-30/3/40	HG
12/5/41-31/5/41	LS
29/12/41-31/1/42	LO
16/11/42-2/1/43	HG
23/6/43-27/7/43	HS
15/11/43-4/12/43	LO
26/12/44-3/2/45	HS
4/3/46-5/4/46	HG
27/9/47-8/11/47	LS
28/9/48-1/11/48	LS
16/10/49-29/11/49	HG
21/3/51-18/4/51	HI
12/8/52-8/9/52	LI
19/10/53-14/11/53	LI
15/9/54-22/10/54	HG
3/8/56-28/8/56	HI
23/5/57-6/6/57	LC(EO)
7/6/58-15/7/58	LI
1/8/60-9/9/60	HG
16/9/60-26/9/60	NCRect(EO)
6/3/63-24/3/63	HI

Boilers

New	8657
25/11/36	9735 New (sloping throatplate)
30/3/40	9443 from 5323 (sloping throatplate)
2/1/43	9549 from 5429 (sloping throatplate)
5/4/46	9507 from 5460 (sloping throatplate)
29/11/49	10346 from 5390 (sloping throatplate)
22/10/54	9381 from 44846 (sloping throatplate)
9/9/60	12230 from 45354 (sloping throatplate)

Tenders

New	9094
?	10544 (part-welded)

Mileage/(weekdays out of service)

1934	15,042 (5)
1935	48,710 (124)
1936	43,720 (138)
1937	47,629 (97)
1938	38,036 (140)
1939	30,149 (121)
1940	52,438 (55)
1941	52,778 (53)
1942	38,860 (75)
1943	49,004 (66)
1944	45,541 (49)
1945	38,791 (98)
1946	49,407 (82)
1947	34,405 (117)
1948	50,879 (59)
1949	42,544 (73)
1950	54,948 (44)
1951	45,173 (51)
1952	34,995 (81)
1953	36,814 (59)
1954	40,922 (72)
1955	35,328 (43)
1956	29,515 (44)
1957	42,132 (73)
1958	43,094 (68)
1959	40,535
1960	25,248

Mileage at 12/36: 107,472
Mileage at 31/12/50: 732,881

Sheds

Crewe	6/10/34
Carlisle W	27/10/34
Nottingham	10/2/35
Leeds	10/2/40
Saltley	30/9/50
Bristol	1/12/56 (loan)
Saltley	13/4/57
Oxley	3/4/65
Crewe South	6/3/67

Withdrawn w.e. 22/7/67

45040 was one of the early sloping throatplate conversions in November 1936. It has the later pattern of top feed cover but has retained its original riveted tender. The picture was taken around 1965, possibly when it was being transferred from Saltley to Oxley, given the question mark chalked where its shedplate should be fixed.

45040 at its home shed Holbeck in one of those interim BR non-liveries. It was renumbered in November 1948 using LMS pattern numbers but its welded tender is devoid of any ownership markings. 45040 had been converted to a sloping throatplate boiler in November 1936.

45041

Built as 5041 at Vulcan Foundry 6/10/34
Renumbered 45041 w.e. 15/1/49

Improvements and modifications

12/6/38	BTH speed indicator
4/1/39	Removal of vacuum pump
?	Steam sanding
19/11/59	Fitting BR ATC equipment

Repairs

11/9/35-11/10/35	LS
9/6/36-30/6/36	LO
4/9/36-15/10/36	LS
8/3/37-28/4/37	HG
2/11/37-6/12/37	HS
4/6/38-13/6/38	LO
17/10/38-4/1/39	HS
17/7/39-9/8/39	LO
20/12/39-3/2/40	HG
27/12/40-25/1/41	HS
1/12/41-3/1/42	LS
30/3/42-18/4/42	LO
15/2/43-5/3/43	LS
5/5/44-27/5/44	HG
9/11/45-8/12/45	HS
16/4/47-13/6/47	HG
22/12/48-14/1/49	LI
14/1/50-6/2/50	LI
16/8/51-15/9/51	HG
20/7/53-18/8/53	HI
24/8/54-1/10/54	HG
10/10/56-9/11/56	HI
8/6/57-28/6/57	LC(EO)
27/1/58-27/2/58	HI
12/10/59-19/11/59	HG
9/10/61-8/11/61	HI
7/5/63-30/5/63	LI
10/5/65-15/6/65	HI
9/7/66-30/7/66	LC

Boilers

New	8658	
16/4/37	8671 from 5054 (domed)	
3/2/40	8987 from 5207	
27/5/44	8823 from 5042 (domed)	
3/6/47	8906 from 5060	
15/9/51	8823 from 45188 (domed)	
1/10/54	8955 from 45031	
19/11/59	8971 from 45034	

Tenders

New	9095

Mileage/(weekdays out of service)

1934	16,703 (2)
1935	55,347 (102)
1936	40,679 (123)
1937	43,548 (117)
1938	32,941 (177)
1939	50,922 (104)
1940	48,101 (68)
1941	47,218 (66)
1942	55,029 (35)
1943	49,226 (51)
1944	42,598 (47)
1945	34,645 (113)
1946	46,955 (56)
1947	36,250 (96)
1948	37,252 (44)
1949	37,335 (54)
1950	35,694 (47)
1951	34,763 (57)
1952	41,951 (36)
1953	34,061 (56)
1954	37,821 (67)
1955	45,080 (34)
1956	32,289 (65)
1957	45,371 (50)
1958	36,591 (79)
1959	23,563
1960	28,228

Mileage at 12/36: 112,729
Mileage at 31/12/50: 710,443

Sheds

Crewe	6/10/34
Carlisle W	27/10/34
Nottingham	10/2/35
Kentish Town	9/7/38
Crewe North	14/6/47 (loan)
Preston	21/6/47 (loan)
Preston	5/7/47
Crewe South	11/10/47
Willesden	28/5/49
Crewe South	18/11/50
Carlisle Upperby	9/2/57
Patricroft	14/6/58
Chester (Mid)	20/9/58
Speke Junction	9/1/60
Holyhead	10/6/61
Edge Hill	28/10/61
Bank Hall	19/10/63
Warrington	21/11/64
Lostock Hall	7/10/67

Withdrawn w.e. 9/12/67

5041 in original condition at Kentish Town while shedded at Nottingham between February 1935 and July 1938. All of the peculiarities of the first Vulcan Foundry engines are well shown including the open front running plate, the absence of steam heat pipes and the worksplate above the steam pipe.

45041 from Crewe South passing through Kings Langley on 25 September 1953 with a mixed freight. It alternated between domed and domeless boilers up until 1959, this being one of the domed periods. However it kept its first riveted tender until withdrawal in 1967.

45042

Built as 5042 at Vulcan Foundry 6/10/34
Renumbered 45042 w.e 2/4/49

Improvements and modifications

2/3/38	Removal of vacuum pump
12/6/38	BTH speed indicator
11/4/44	Steam sanding
2/3/57	Modernisation
6/2/59	Fitting BR ATC equipment

Repairs

29/8/35-15/10/35	LS
21/10/35-4/11/35	LO
26/10/36-7/12/36	LS
12/3/37-2/4/37	HS
10/8/37-30/8/37	LO
9/2/38-2/3/38	HG
15/3/39-21/4/39	HS
9/4/40-25/4/40	HS
20/12/41-24/1/42	LS
14/4/42-7/5/42	LO
14/12/42-5/1/43	LS
5/11/43-20/11/43	LO
27/3/44-11/4/44	HG
27/4/45-16/5/45	HS
16/12/46-8/1/47	LS
9/9/47-10/10/47	HO
3/12/47-24/12/47	LO
16/2/49-31/3/49	HG
16/5/50-2/6/50	LI
15/10/51-9/11/51	HI
30/6/53-12/8/53	HG
22/1/55-12/2/55	LI
25/1/57-2/3/57	HG
16/1/59-6/2/59	LI

Boilers

New	8659
17/2/38	8647 from 5030 (domed)
25/4/40	8823 from 5068 (domed)
11/4/44	9018 from 5217
31/3/49	8660 from 5147 (domed)
12/8/53	8993 from 45051
2/3/57	8981 from 45067

Tenders

New	9096

Mileage/(weekdays out of service)

Year	Mileage (days)
1934	14,323 (6)
1935	60,328 (118)
1936	47,729 (149)
1937	47,897 (141)
1938	50,911 (76)
1939	48,013 (96)
1940	33,429 (56)
1941	25,674 (96)
1942	35,666 (79)
1943	44,177 (46)
1944	46,941 (56)
1945	40,451 (65)
1946	42,232 (58)
1947	35,927 (108)
1948	43,380 (58)
1949	35,659 (60)
1950	36,443 (38)
1951	29,902 (59)
1952	45,121 (24)
1953	35,071 (69)
1954	39,966 (39)
1955	34,861 (81)
1956	30,749 (34)
1957	35,380 (59)
1958	36,522 (29)
1959	39,101
1960	35,644

Mileage at 12/36: 122,380
Mileage at 31/12/50: 689,180

Sheds

Crewe	6/10/34
Carlisle W	27/10/34
Trafford Park	10/2/35
Derby	9/7/38
Toton	15/6/40
Shrewsbury	4/7/42 (loan)
Shrewsbury	15/8/42
Mold Junction	19/5/45
Chester	7/7/45
Patricroft	11/12/48
Llandudno Jct	11/12/48 (loan)
Patricroft	26/2/49
Chester (Mid)	20/9/58
Holyhead	20/6/59
Mold Jct	19/9/59
Holyhead	22/6/63
Mold Junction	14/9/63
Llandudno Jct	23/4/66
Chester	8/10/66
Crewe South	10/6/67

Stored
8/5/67-15/9/67

Withdrawn w.e. 16/9/67

Below. 5042 as delivered at Crewe North in October 1934. It hasn't yet been fitted with a shedplate which it would receive when allocated to 'Carlisle West' (Upperby) by the end of the month. All the original Vulcan Foundry features are apparent including tall chimney, open front framing, hollow bogie axles, scalloped cut-outs on the steam pipes and prominent top feed pipes.

45043

Built as 5043 at Vulcan Foundry 13/10/34
Renumbered 45043 w.e. 24/9/49

Improvements and modifications

20/3/38	BTH speed indicator
29/7/38	Removal of vacuum pump
24/10/42	Steam sanding
24/9/55	Modernisation
31/10/59	Fitting BR ATC equipment

Repairs

30/10/35-4/12/35	LS
2/9/36-14/12/36	LO
30/11/37-3/1/38	HG
4/7/38-29/7/38	LO
15/2/39-7/3/39	LS
18/4/39-9/5/39	LO
15/5/39-2/6/39	LO
6/12/39-3/1/40	HS
24/12/40-23/1/41	LO
14/1/42-14/2/42	LS
26/9/42-24/10/42	HG
8/6/43-30/6/43	LOI
8/2/44-3/3/44	LS
21/5/45-19/7/45	HG
11/9/46-18/10/46	HS
27/2/48-10/4/48	HS
2/8/49-23/9/49	HG
22/2/51-4/4/51	LI
14/4/51-17/4/51	NC Rect
4/6/51-25/6/51	NCLS
23/7/51-17/8/51	LC
26/10/51-31/10/51	NC
3/2/53-7/3/53	HI
1/2/54-5/3/54	HI
25/8/55-24/9/55	G
11/3/58-2/4/58	LI
13/10/59-31/10/59	
31/10/60-17/12/60	LI
16/2/62-13/3/62	HC
8/3/65-27/3/65	INT

Boilers

New	8660
15/12/37	8640 from 5023 (domed)
24/10/42	8949 from 5174
19/7/45	8951 from 5208
23/9/49	8964 from 5051
24/9/55	9036 from 45092
?	9007

Tenders

New	9097
7/3/53	10548 (part-welded)

Mileage/(weekdays out of service)

1934	10,792 (16)
1935	50,373 (91)
1936	36,688 (160)
1937	57,113 (109)
1938	61,367 (85)
1939	42,371 (116)
1940	32,445 (28)
1941	23,506 (89)
1942	50,813 (63)
1943	43,118 (61)
1944	45,813 (62)
1945	39,478 (113)
1946	39,458 (110)
1947	48,132 (55)
1948	45,630 (65)
1949	40,791 (69)
1950	37,578 (47)
1951	30,785 (102)
1952	38,148 (63)
1953	41,369 (57)
1954	40,434 (82)
1955	31,105 (60)
1956	41,081 (31)
1957	38,565 (49)
1958	36,647 (48)
1959	35,505
1960	33,574

Mileage at 12/36: 97,853
Mileage at 31/12/50: 705,466

Sheds

Crewe	13/10/34
Edge Hill	27/10/34
Bristol	23/2/35
Derby	12/3/38
Normanton	16/3/40
Leeds	18/10/41
Low Moor	10/7/48 (loan)
Low Moor	14/8/48
Blackpool	10/6/50
Perth	8/9/51 (loan)
Perth	15/9/51
Carlisle Kingmoor	17/5/52
St Margarets	25/7/53
Chester	19/9/53
Mold Jct	24/9/55 (loan)
Mold Jct	15/9/56
Rugby	10/1/59
Mold Jct	17/1/59
Holyhead	22/6/63
Mold Jct	14/9/63
Holyhead	23/4/66
Speke Junction	10/12/66

Stored
11/7/66-28/11/66

Withdrawn w.e. 25/11/67

45043 fresh off Crewe Works and in immaculate LNWR style lined black. It was at Bradford Low Moor until June 1950 – it is assumed that the period is September 1949 – after completion of a Heavy General overhaul. It remained domeless apart from a short spell following its first boiler change in 1937 and also kept its original riveted tender until March 1953 when it changed to a part-welded type.

45044 resting at Willesden on 28 April 1963. It had carried a domeless boiler since December 1950, was paired with a welded tender from April 1953 and received AWS in May 1959. When this picture was taken it was on loan from Willesden to Northampton.

45044

Built as 5044 at Vulcan Foundry 13/10/34
Renumbered 45044 w.e. 19/6/48

Improvements and modifications
21/7/38	Removal of vacuum pump
18/1/40	BTH speed indicator
30/10/43	Steam sanding
21/4/59	Fitting BR ATC equipment

Repairs
7/11/35-16/12/35	LS
18/11/36-7/12/36	LS
5/4/37-16/4/37	LO
8/5/37-11/6/37	HG
30/6/38-21/7/38	LS
28/12/39-18/1/40	HG
30/1/41-15/2/41	HS
31/8/42-16/9/42	LS
12/10/43-30/10/43	HG
29/2/44-6/3/44	LO
27/4/45-26/5/45	LS
4/6/46-17/7/46	LO
24/7/47-8/9/47	HG
25/5/48-15/6/48	LO
4/10/49-4/11/49	LI
6/11/50-9/12/50	HG
4/11/51-28/11/51	LI
1/12/52-24/12/52	LC
4/1/54-23/1/54	HI
4/12/55-31/12/55	HG
31/10/57-30/11/57	LI
22/3/59-21/4/59	LI
18/8/59-1/9/59	LC(EO)
12/4/60-13/5/60	LC(EO)
25/1/61-25/2/61	HG
12/4/62-9/5/62	INT
11/2/65-27/2/65	Rect
20/5/65-31/5/65	LC
7/2/66-11/3/66	Collision damage

Boilers
New	8661
20/5/37	8643 from 5026 (domed)
18/1/40	9001 from 5221
30/10/43	9019 from 5053
8/9/47	8675 from 5176 (domed)
9/12/50	8992 from 5145
31/12/55	9019 from 45191
25/2/61	9042 from 45135

Tenders
New	9098
7/4/53	9236 (welded)

Mileage/(weekdays out of service)
1934	12,087 (11)
1935	50,442 (110)
1936	48,133 (102)
1937	52,391 (80)
1938	44,659 (64)
1939	40,702 (39)
1940	42,191 (47)
1941	37,592 (64)
1942	28,890 (44)
1943	35,604 (38)
1944	35,667 (32)
1945	32,366 (66)
1946	26,556 (70)
1947	29,933 (73)
1948	31,077 (68)
1949	34,430 (69)
1950	32,884 (61)
1951	38,737 (38)
1952	38,763 (62)
1953	40,352 (30)
1954	39,867 (47)
1955	37,221 (65)
1956	43,872 (21)
1957	33,443 (54)
1958	41,681 (44)
1959	30,193
1960	36,750

Mileage at 12/36: 110,662
Mileage at 31/12/50: 615,604

Sheds
Crewe	12/10/34
Edge Hill	27/10/34
Bushbury	?
Edge Hill	1/2/35
Midland Division	14/2/35 (loan)
Bristol	1/4/35
Patricroft	7/8/37
Preston	6/7/40
Edge Hill	27/3/43
Crewe South	29/9/45
Stoke	20/6/59
Longsight	29/10/60
Carlisle Upperby	19/11/60 (loan)
Longsight	31/3/62
Willesden	31/3/62
Northampton	9/2/63 (loan)
Rugby	17/8/63
Willesden	26/10/63
Chester (M)	2/11/63

Withdrawn w.e. 5/11/66

45045

Built as 5045 at Vulcan Foundry 13/10/34
Renumbered 45045 w.e. 10/4/48

Improvements and modifications

10/6/38	Removal of vacuum pump
?	Steam sanding
8/11/54	Sloping throatplate boiler
13/6/59	Fitting BR ATC equipment

Repairs

6/9/35-25/9/35	LS
16/8/36-3/9/36	LO
9/11/36-23/11/36	LS
14/4/37-29/4/37	LO
16/4/38-10/6/38	HG
5/1/40-2/2/40	LS
13/5/41-29/5/41	HS
7/2/42-28/2/42	LO
14/8/42-12/9/42	LS
4/10/43-16/10/43	HG
21/8/44-2/9/44	LO
5/2/45-15/2/45	LO
27/8/45-22/9/45	LS
7/2/47-13/3/47	LS
17/3/48-9/4/48	LO
15/8/48-9/9/48	HG
15/8/50-2/9/50	HI
23/11/51-11/1/52	LI
17/2/53-18/3/53	HG
21/11/53-18/12/53	LC(EO)
24/2/54-13/3/54	LC(EO)
9/10/54-8/11/54	HG
10/2/56-8/3/56	HI
9/5/57-1/6/57	LI
4/5/59-13/6/59	HI
22/6/59-25/6/59	NCRect (EO)
7/11/59-12/12/59	HC(EO)
13/5/61-10/8/61	LI
18/10/62-10/11/62	LC
20/1/64-15/2/64	HI

Boilers

New	8662
24/5/38	9025 from 5095
16/10/43	8927 from 5006
9/9/48	8921 from 5004
18/3/53	8681 from 45114 (domed)
8/11/54	9471 from 44847 (sloping throatplate)
12/12/59	9474 from 45437 (sloping throatplate)

Tenders

New	9099
10/8/61	10446 (welded)
10/10/62	10726 (part-welded)
26/10/66	9488 (welded)

Mileage/(weekdays out of service)

1934	14,450 (4)
1935	42,892 (60)
1936	44,875 (79)
1937	44,184 (59)
1938	38,107 (112)
1939	51,776 (43)
1940	41,215 (45)
1941	38,433 (42)
1942	29,990 (68)
1943	36,973 (32)
1944	35,446 (36)
1945	29,044 (66)
1946	31,050 (42)
1947	26,692 (61)
1948	34,478 (73)
1949	34,260 (38)
1950	38,570 (65)
1951	36,877 (73)
1952	44,350 (55)
1953	43,549 (83)
1954	42,369 (71)
1955	45,621 (48)
1956	37,120 (56)
1957	37,888 (48)
1958	31,497 (40)
1959	29,492
1960	40,238

Mileage at 12/36: 102,217
Mileage at 31/12/50: 612,435

Sheds

Crewe	13/10/34
Edge Hill	27/10/34
Carlisle	2/3/35 (loan)
Edge Hill	30/3/35
Stoke	27/4/35
Crewe	4/5/35
Stoke	8/6/35
Shrewsbury	9/11/35
Llandudno Jct	3/7/37
Carlisle W	18/1/38
Mold Jct	22/1/38 (loan)
Carlisle W	5/2/38
Carnforth	4/5/40
Edge Hill	25/10/41
Crewe South	9/3/46 (loan)
Crewe South	4/5/46
Chester	16/8/47
Edge Hill	25/10/47
Llandudno Jct	5/6/48
Carlisle Upperby	26/6/48
Crewe North	25/12/48
Springs Branch	8/1/49
Chester	28/5/49
Holyhead	7/7/51
Crewe South	15/9/56
Crewe North	31/12/60
Crewe South	6/1/62
Holyhead	8/12/62
Llandudno Jct	2/3/63
Shrewsbury	2/10/65
Croes Newydd	27/11/65
Heaton Mersey	6/8/66 (on loan)
Heaton Mersey	20/8/66

Stored
23/2/66-18/3/66

Withdrawn w.e. 29/10/66

Llandudno Junction's 45045 on 16 May 1964 looks as if it has just picked up water from the troughs at Prestatyn. It was a late conversion to a sloping throatplate boiler, in November 1954. It acquired the part-welded tender in October 1962 and the AWS in June 1959. Photograph J. Hobbs.

With plenty of steam by the look of it, Crewe South's 45046 is a long way from home at the Waterloo bufferstops. It spent much of the summer of 1965 on the Southern Region and was noted on the 21.24 Bournemouth to Eastleigh local on 16 June and then the 15.35 Waterloo-Bournemouth throughout the next week. 45046 was used on a number of occasions until the end of the month on a diagram covering the 08.01 Eastleigh-Waterloo and 15.35 Waterloo-Bournemouth. On Saturday 26 June 4 it arrived in Bournemouth with a train from Leeds but was back in Waterloo again on 28 August. Photograph www.rail-online.co.uk

45046

Built as 5046 at Vulcan Foundry 20/10/34
Renumbered 45046 w.e. 23/4/49

Improvements and modifications

22/4/39	Removal of vacuum pump
17/7/39	BTH speed indicator
29/8/45	Steam sanding
11/6/59	Fitting BR ATC equipment

Repairs

1/8/35-15/8/35	LO
5/5/36-22/5/36	LS
15/3/37-12/4/37	HO
28/12/37-17/1/38	HG
12/6/39-17/7/39	HS
6/2/41-28/2/41	HG
5/8/42-3/9/42	LS
11/6/43-24/6/43	LS
10/8/45-29/8/45	HG
29/7/47-6/9/47	HS
28/3/49-20/4/49	HG
31/10/49-16/11/49	LC
20/7/50-28/8/50	HI
8/9/51-26/10/51	HI
22/4/53-22/5/53	HG
6/4/55-5/5/55	LI
5/12/55-7/1/56	LC(EO)
21/6/57-3/8/57	HG
16/5/59-11/6/59	HI
21/5/60-25/6/60	HI
14/11/61-16/12/61	LI
31/7/62-23/8/62	HC
5/4/65-6/5/65	LI

Boilers

New	8663
4/1/38	8654 from 5037 (domed)
28/2/41	8653 from 5092 (domed)
29/8/45	8930 from 5059
20/4/49	8683 from 5000 (domed)
22/5/53	8676 from 45118 (domed)
3/8/57	8665 from 45014 (domed)
?	9036

Tenders

New	9100

Mileage at 12/36: 110,207
Mileage at 31/12/50: 651,543

Sheds

Crewe	20/10/34
Edge Hill	27/10/34
Carlisle W	2/3/35
Edge Hill	30/3/35
Stoke	27/4/35
Crewe	4/5/35
Stoke	8/6/35
Crewe North	24/9/38
Monument Lane	4/3/39
Aston	30/9/39
Rugby	13/9/41
Willesden	21/3/42
Crewe North	23/10/42
Crewe South	16/1/43
Preston	6/2/43
Carlisle U	15/11/47 (loan)
Preston	7/2/48
Barrow	14/2/48
Bescot	12/5/56
Patricroft	9/6/56
Carnforth	15/9/56
Longsight	7/11/59
Crewe North	9/1/60
Crewe South	23/6/62
Crewe North	15/9/62
Crewe South	14/9/63
Stockport	21/8/65
Bolton	11/5/68

Stored
7/11/66-28/10/67

Withdrawn w.e. 22/6/68

Mileage/(weekdays out of service)

1934	12,737 (5)
1935	46,559 (42)
1936	50,911 (41)
1937	39,416 (71)
1938	52,062 (49)
1939	41,527 (62)
1940	34,434 (44)
1941	34,717 (41)
1942	35,363 (76)
1943	40,683 (35)
1944	31,573 (46)
1945	33,467 (49)
1946	35,334 (53)
1947	30,648 (70)
1948	44,466 (29)
1949	39,774 (64)
1950	47,872 (56)
1951	41,259 (74)
1952	52,375 (28)
1953	47,627 (57)
1954	42,465 (52)
1955	37,317 (63)
1956	35,686 (52)
1957	37,410 (77)
1958	31,250 (51)
1959	36,535
1960	39,078

Below. 45046 at Stockport in 1966 with a crudely painted 9B Edgeley shedplate. It had a domeless boiler from August 1962 although it had spent most of its life with the domed variety and the riveted tender was the one it had had from new. Too much ash from the smokebox, AWS and a lowered top lamp bracket complete the picture.

45047

Built as 5047 at Vulcan Foundry 20/10/34
Renumbered 45047 w.e. 14/8/48

Improvements and modifications

31/12/36	Sloping throatplate boiler
15/6/38	Removal of vacuum pump
7/7/47	Steam sanding
?	Fitting BR ATC equipment

Repairs

6/11/35-25/11/35	LS
28/11/36-13/1/37	HG
13/5/38-15/6/38	HS
26/1/39-27/2/39	LS
8/9/39-17/10/39	LO
15/7/40-7/8/40	HG
13/8/41-20/9/41	HS
17/4/42-22/5/42	LS
14/1/43-3/2/43	LS
26/11/43-20/12/43	LS
17/7/44-14/8/44	HS
29/9/45-2/11/45	HG
25/5/46-22/6/46	LO
18/10/46-30/11/46	LO
24/5/47-7/7/47	LS
23/6/48-12/8/48	HS
30/11/48-5/1/49	LC
24/3/49-27/4/49	HI
26/12/49-29/12/49	LC
13/1/50-26/1/50	LC
23/5/50-24/5/50	TO
14/8/50-7/10/50	G
21/11/50-18/1/51	LC
7/11/51-1/12/51	HI
24/3/52-11/4/52	NC(EO)
21/8/52-4/10/52	HI
5/10/53-7/11/53	LI(EO)
27/11/53-27/11/53	HC(TO)
1/8/55-24/9/55	G
16/12/55-28/12/55	LC(EO)
9/4/57-27/5/57	HI
27/9/58-23/10/58	LC(EO)
1/9/59-16/10/59	G
24/3/60-8/4/60	LC(EO)
17/5/61-22/6/61	HI
30/5/63-4/7/63	HI
9/7/63-10/7/63	NC
1/3/65-6/3/65	LC

Boilers

New	8664
31/12/36	9734 (sloping throatplate)
7/8/40	9485 from 5365 (sloping throatplate)
2/11/45	9489 from 5423 (sloping throatplate)
7/10/50	11321 from 5453 (sloping throatplate)

Tenders

New	9101
10/10/46	9279 (welded)
6/6/47	9716 (welded)
7/10/50	9069
?	10806 (part-welded)
7/11/53	10689 (part-welded)
?	10504 (welded)
23/9/55	9298 (welded)

Mileage/(weekdays out of service)

1934	9,215 (1)
1935	65,234 (48)
1936	52,589 (66)
1937	51,834 (76)
1938	46,445 (102)
1939	35,100 (116)
1940	28,691 (66)
1941	33,052 (85)
1942	67,396 (48)
1943	53,540 (61)
1944	39,555 (98)
1945	43,916 (83)
1946	24,877 (160)
1947	38,067 (129)
1948	40,477 (99)
1949	48,099 (77)
1950	24,452 (160)

Mileage at 12/36: 127,038
Mileage at 31/12/50: 702,539

Sheds

Edge Hill	27/10/34
Carlisle Kingmoor	29/11/34
Crewe	25/11/35
Blackpool	26/6/37
Aintree	2/11/40
Agecroft	25/1/41
Blackpool	5/4/41
Accrington	12/7/41
Perth	27/12/41
Corkerhill	26/2/44
Carlisle N	12/5/45
Corkerhill	6/10/45
Ayr	27/4/52
Fort William	24/5/54
Perth	15/11/54
Dalry Road	12/9/65
St Margarets	4/10/65

Stored
13/3/39-3/4/39
17/4/39-8/5/39

Withdrawn 17/7/66

5047 on 17 coaches of very mixed parentage from LNWR, LYR and MR origins, probably ECS. The presence of a Scottish loco behind suggests a location north of the border, probably when it was allocated to Upperby from November 1934 to November 1935 because it still has its original boiler with prominent top feed pipes and tall chimney which it lost at the end of 1936 when it received a sloping throatplate boiler.

45048 in final condition at Willesden on 9 May 1964 had been domed since July 1956; the riveted tender is the original one. It was allocated to Crewe South at this date but moved to the North shed in June, having previously bounced back and forth between the two sheds on two previous occasions.

45048

Built as 5048 at Vulcan Foundry 20/10/34
Renumbered 45048 w.e. 21/1/50

Improvements and modifications
22/9/38	Removal of vacuum pump
8/2/44	Steam sanding
20/7/56	Modernisation
16/5/59	Fitting BR ATC equipment

Repairs
28/10/35-10/12/35	LS
8/1/36-17/2/36	LO
25/7/36-22/8/36	LO
15/6/37-30/6/37	HS
23/8/38-22/9/38	HG
14/2/40-16/3/40	HS
18/3/42-11/4/42	LS
22/10/42-20/11/42	LS
11/1/44-8/2/44	HG
13/9/44-6/10/44	TRO
10/4/46-3/5/46	HS
13/1/48-19/2/48	HG
2/1/50-20/1/50	HI
11/12/51-19/1/52	HG
12/10/53-5/11/53	HI
24/3/55-22/4/55	LI
31/5/56-20/7/56	HG
3/4/58-23/5/58	HI
7/5/59-16/5/59	NC(EO)
25/5/59-12/6/59	LC
2/8/59-12/9/59	HC
14/12/60-20/1/61	HG
23/9/63-28/10/63	HI

Boilers
New	8665
22/9/38	8911 from 5131
8/2/44	8679 from 5098 (domed)
19/2/48	9035 from 5105
19/1/52	8906 from 45041
20/7/56	8655 from 45193 (domed)
20/1/61	8823 from 45069 (domed)

Tenders
New	9102
3/8/66	10498 (welded)

Mileage/(weekdays out of service)
1934	9,958 (6)
1935	52,612 (68)
1936	50,398 (76)
1937	42,095 (62)
1938	43,374 (83)
1939	49,646 (74)
1940	28,208 (59)
1941	27,841 (54)
1942	32,441 (70)
1943	26,460 (41)
1944	25,496 (87)
1945	33,492 (49)
1946	30,692 (45)
1947	30,759 (48)
1948	33,190 (74)
1949	30,638 (35)
1950	37,100 (61)
1951	38,441 (46)
1952	44,982 (40)
1953	35,148 (56)
1954	41,420 (33)
1955	37,343 (67)
1956	36,665 (78)
1957	36,105 (53)
1958	36,505 (75)
1959	33,111
1960	39,710

Mileage at 12/36: 112,968
Mileage at 31/12/50: 584,400

Sheds
Crewe	22/10/34
Edge Hill	27/10/34
Sheffield	8/12/34
Llandudno Jct	14/11/36
Crewe South	6/4/40
Crewe North	26/6/43
Crewe South	7/8/43
Willesden	5/6/48
Crewe South	23/10/48
Holyhead	5/7/52
Crewe North	11/10/52
Crewe South	18/4/53
Rugby	21/4/56 (loan)
Crewe South	9/6/56
Crewe North	20/6/64
Bescot	19/9/64
Stourbridge	26/3/66
Holyhead	28/5/66
Springs Branch	16/12/66

Stored
12/1/66-28/5/66
1/8/66-2/12/66

Withdrawn w.e. 2/12/67

45049

Built as 5049 at Vulcan Foundry 27/10/34
Renumbered 45049 w.e. 1/5/48

Improvements and modifications
22/6/38	Removal of vacuum pump
29/5/43	Steam sanding
16/7/54	Sloping throatplate boiler
16/7/54	Modernisation
7/8/59	Fitting BR ATC equipment

Repairs
25/11/35-16/12/35	LS
1/10/36-22/10/36	LS
6/5/38-22/6/38	HG
3/1/40-20/1/40	HS
31/1/41-14/2/41	HS
15/5/42-5/6/42	LS
6/5/43-29/5/43	LS
11/1/44-10/2/44	LS
10/7/44-5/8/44	LO
15/5/45-14/6/45	LS
8/2/46-9/3/46	HS
18/12/46-14/2/47	HG
30/10/47-12/11/47	LO
20/3/48-1/5/48	LS
19/7/48-29/7/48	LO
25/10/48-6/11/48	LO
9/12/48-14/1/49	LC
7/3/49-7/4/49	HI
11/4/49-13/4/49	NC(R)
24/8/49-2/9/49	LC
27/10/49-24/11/49	LC
30/11/49-3/12/49	NC(R)
13/2/50-18/3/50	HI
3/4/50-8/4/50	NC(R)
25/8/50-4/10/50	LC
4/1/51-21/4/51	G
22/2/52-20/3/52	HI
10/2/53-14/3/53	HI
7/5/54-16/7/54	G
24/2/55-9/3/55	LC(EO)
14/3/55-19/3/55	LC(EO)
1/8/55-12/8/55	LC
1/2/56-25/2/56	HI
27/8/56-31/8/56	LC(EO)
13/2/57-2/3/57	LC(EO)
3/8/57-22/8/57	HI
23/6/59-7/8/59	G
5/4/60-14/4/60	LC
31/7/61-16/9/61	LI

Boilers
New	8666
7/6/38	9022 from 5092
29/5/43	8645 from 5119 (domed)
14/2/47	8976 from 5009
21/4/51	8653 from 45153 (domed)
16/7/54	12442 from 44903 (sloping throatplate)
7/8/59	8820 from 45081 (domed)

Tenders
New	9103
21/5/45	9175
12/1/49	10527 (welded)
17/3/50	10686 (part-welded)
8/1/51	9164
5/4/51	10618 (part-welded)
31/8/51	10698 (part-welded)
19/3/52	10620 (part-welded)
13/3/53	9841 (welded)
9/7/54	9502 (welded)
25/2/56	9120
23/9/61	10679 (part-welded)
22/8/63	10547 (part-welded)

Mileage/(weekdays out of service)
1934	12,286 (-)
1935	56,442 (58)
1936	55,875 (40)
1937	45,627 (71)
1938	36,449 (156)
1939	36,315
1940	27,812 (62)
1941	35,382 (52)
1942	37,461 (48)
1943	56,011 (41)
1944	47,413 (79)
1945	50,062 (44)
1946	44,327 (65)
1947	43,207 (88)
1948	35,085 (114)
1949	43,086 (105)
1950	38,469 (118)
1951	32,571 (147)
1952	39,761 (76)
1953	37,493 (79)
1954	40,323 (93)
1955	44,328 (77)
1956	44,836 (94)
1957	36,883 (70)
1958	38,814 (49)
1959	33,369
1960	41,751
1961	29,077
1962	32,208

Mileage at 12/36: 124,603
Mileage at 31/12/50: 701,309

Sheds
Crewe	3/11/34
Edge Hill	2/2/35
Carlisle W	2/3/35
Crewe	28/9/35
Farnley Jct	26/6/37
Agecroft	17/1/42
Perth	22/11/42
Corkerhill	26/2/44
Carlisle N	12/5/45
Corkerhill	6/10/45
Ayr	3/5/52
Fort William	16/7/54
Perth	15/11/54
Stirling	20/8/56
Perth	25/8/56
Stirling	15/2/57

Withdrawn w.e. 29/8/63

5049 shows the original LMS livery with 12in cab numbers, 5P 5F, lining around the four edges of the cab side sheet and below the cab windows, and the closer 40in spacing of the 15in letters on the tender which was characteristic of the locomotives built by 'the Trade'. Note the strengthening webs and hollowed out axles on the coupled wheels, gravity sanding, the plain tender axlebox covers and the crosshead driven vacuum pump which was removed in June 1938. Photograph W. Hermiston, www.transporttreasury.co.uk

45050

Built as 5050 at Vulcan Foundry 3/11/34
Renumbered 45050 w.e. 1/10/49

Improvements and modifications

2/2/39	Removal of vacuum pump
8/5/43	Steam sanding
20/3/59	Fitting BR ATC equipment
20/4/63	Fitting Smith-Stone speedometer

Repairs

11/3/36-1/4/36	LS
20/10/36-12/11/36	LO
20/4/37-2/6/37	HG
9/1/39-2/2/39	LS
29/7/40-21/8/40	HG
24/2/41-14/3/41	HO
23/10/41-8/11/41	LS
23/4/43-8/5/43	HS
2/8/44-16/8/44	LS
18/11/44-16/12/44	LO
10/12/45-12/1/46	LS
12/10/46-1/11/46	LO
25/10/47-13/12/47	HG
3/9/49-29/9/49	HI
14/6/50-30/6/50	LI
2/1/52-2/2/52	HG
5/6/52-28/6/52	LC
26/1/53-21/2/53	HI
10/3/53-20/3/53	NC(Rect)
12/5/54-3/6/54	HI
17/11/55-16/12/55	HG
7/1/57-31/1/57	LI
7/12/57-20/12/57	LC(EO)
8/8/58-4/9/58	LI
31/1/59-20/3/59	LC(EO)
16/5/60-24/6/60	HG
12/8/60-19/8/60	NC(Rect EO)
18/4/61-1/6/61	LC(EO)
27/7/62-18/8/62	HI
30/4/65-25/5/65	INT

Boilers

New	8667
17/5/37	9027 from 5097
21/8/40	8083 from 5203
14/3/41	9023 from 5213
8/5/43	9053 from 5060
13/12/47	8680 from 5201 (domed)
2/2/52	8637 from 45198 (domed)
16/12/55	9047 from 45210
?	9027

Tenders

New	9104
26/1/53	10534 (welded)
11/8/62	9496 (welded)

Mileage/(weekdays out of service)

1934	4,423 (18)
1935	57,682 (73)
1936	49,865 (125)
1937	45,130 (56)
1938	49,283 (34)
1939	48,775 (54)
1940	29,794 (72)
1941	34,647 (73)
1942	39,368 (19)
1943	33,517 (38)
1944	33,012 (60)
1945	36,413 (40)
1946	35,441 (44)
1947	36,101 (75)
1948	48,104 (26)
1949	43,552 (45)
1950	49,228 (36)
1951	47,220 (27)
1952	38,929 (72)
1953	38,887 (62)
1954	38,600 (49)
1955	31,121 (66)
1956	43,280 (38)
1957	40,901 (62)
1958	34,553 (59)
1959	39,012
1960	28,184

Mileage at 12/36: 111,970
Mileage at 31/12/50: 674,335

Sheds

Crewe	10/11/34
Bushbury	2/2/35
Saltley	16/2/35 (loan)
Saltley	6/4/35
Derby	27/4/35
Edge Hill	29/11/36
Carlisle W	22/4/39
Patricroft	26/10/40
Carlisle W	11/4/42
Carnforth	2/5/42
Carlisle Upperby	5/7/52
Rugby	10/1/53
Northampton	15/9/56
Willesden	9/11/60
Warrington	14/1/61
Llandudno Jct	1/7/61
Rugby	23/9/61
Northampton	17/11/62
Chester	20/6/64
Stoke	19/9/64

Stored
17/4/67

Withdrawn w.e. 8/12/67

45050 drifts through Kings Langley on 27 August 1960. It got its first welded tender in 1953, its sixth domeless boiler in late-1955 and had been allocated to Northampton since 1956, receiving AWS in March 1959.

45050 at Willesden during its time at Carnforth, before moving to Upperby in June 1952. It had been renumbered in September 1949 on completion of a Heavy Intermediate repair when it was repainted in BR lined mixed traffic livery. It carried two domed boilers between December 1947 and December 1955 and kept its original riveted tender until January 1953.

45051

Built as 5051 at Vulcan Foundry 10/11/34
Renumbered 45051 w.e. 4/6/49

Improvements and modifications

7/9/38	Removal of vacuum pump
26/4/44	Steam sanding
2/5/59	Fitting BR ATC equipment

Repairs

5/4/35-7/5/35	LO
17/8/35-9/9/35	LO
28/10/35-14/11/35	LO
5/2/36-17/2/36	LO
20/4/36-5/5/36	LS
3/3/37-8/4/37	HG
22/8/38-7/9/38	LS
21/11/39-16/12/39	HG
12/8/41-6/9/41	LS
11/12/42-7/1/43	LS
8/4/44-26/4/44	HG
10/12/45-5/1/46	HS
8/1/47-1/3/47	LO
24/2/48-23/3/48	LS
2/5/49-31/5/49	HG
14/10/50-9/11/50	LI
10/11/51-12/12/51	HI
13/3/53-17/4/53	HG
23/8/54-22/9/54	HI
21/2/55-17/3/55	LC(EO)
13/8/55-3/9/55	LC(EO)
26/3/56-6/4/56	LC(EO)
15/9/56-18/10/56	HI
3/6/57-21/6/57	LC(EO)
27/8/57-26/9/57	LC(EO)
16/6/58-4/7/58	HG
21/4/59-2/5/59	NC(EO)
4/9/59-23/10/59	LI

Boilers

New	8668
17/3/37	8657 from 5040 (domed)
21/11/39	8925 from 5145
26/4/44	8964 from 5086
31/5/49	8993 from 5180
17/4/53	8921 from 45045
4/7/58	8911 from 45184

Tenders

New	9105
8/9/36	9106
28/9/36	9105
5/1/46	9566 (welded)

Mileage/(weekdays out of service)

1934	5,574 (4)
1935	35,002 (111)
1936	56,931 (58)
1937	48,130 (92)
1938	45,243 (70)
1939	39,377 (55)
1940	32,610 (42)
1941	33,205 (58)
1942	31,406 (26)
1943	43,618 (32)
1944	27,389 (46)
1945	30,018 (47)
1946	37,392 (30)
1947	30,976 (69)
1948	35,129 (49)
1949	37,538 (46)
1950	37,061 (52)
1951	34,103 (63)
1952	42,322 (66)
1953	41,793 (71)
1954	33,591 (84)
1955	34,050 (91)
1956	26,907 (90)
1957	30,026 (64)
1958	36,234 (38)
1959	32,561
1960	43,464

Mileage at 12/36: 97,507
Mileage at 31/12/50: 606,599

Sheds

Crewe	17/11/34
Bushbury	2/2/35
Crewe	20/4/35
Willesden	25/9/37
Monument Lane	14/5/38
Patricroft	6/8/38
Llandudno Jct	13/8/38
Aston	24/9/38
Monument Lane	7/7/51
Southern Region	23/5/53
Monument Lane	27/6/53
Bescot	28/4/56
Aston	20/6/59
Monument Lane	1/7/61
Bescot	2/9/61
Lancaster	14/10/61
Northampton	13/1/62
Tyseley	2/10/65
Shrewsbury	12/11/66

Withdrawn w.e. 19/11/66

5051 from Aston piloting a Jubilee in the late 1940s. It has high positioned late LMS 1946 pattern cab numbers and a welded tender, which it acquired in January 1946.

45051 with 2H painted on the smokebox door and Northampton on the bufferbeam. It was shedded there from January 1962 until transferred to Tyseley in October 1965. The picture cannot be dated exactly but the lowered top lamp iron suggests 1964/65. It had been domeless since 1939, had a welded tender from 1946 and was fitted with AWS in May 1959. The steam lance cock cover is also missing adding to the air of neglect. Photograph www.rail-online.co.uk.

45052

Built as 5052 at Vulcan Foundry 17/11/34
Renumbered 45052 w.e. 15/5/48

Improvements and modifications
24/6/38	Removal of vacuum pump
31/1/42	Steam sanding
5/4/56	Modification
2/5/59	Fitting BR ATC equipment

Repairs
24/10/35-14/12/35	LS
18/7/36-18/8/36	LS
20/4/37-6/5/37	LO
11/8/37-25/8/37	LO
2/5/38-24/6/38	HG
4/4/40-18/4/40	LS
13/1/42-31/1/42	HG
16/4/43-6/5/43	LS
20/7/44-5/8/44	LS
10/9/45-6/10/45	HG
4/6/47-21/7/47	LS
6/4/48-12/5/48	LO
7/4/49-9/5/49	HI
8/9/50-5/10/50	HG
12/11/51-15/12/51	LI
21/5/53-16/6/53	LI
19/3/54-31/3/54	LC(EO)
21/6/54-29/7/54	LI
2/2/55-11/2/55	LC(EO)
13/4/55-17/5/55	HC(EO)
3/3/56-5/4/56	HG
1/2/58-21/2/58	LI
2/4/59-2/5/59	LC(EO)
8/5/59-21/5/59	LC
4/3/61-12/4/61	HI

Boilers
New	8669
7/6/38	9043 from 5113
31/1/42	8981 from 5028
6/10/45	8639 from 5206 (domed)
5/10/50	8654 from 5139 (domed)
5/4/56	8985 from 45203

Tenders
New	9106
8/9/36	9105
28/9/36	9106
23/6/54	9088

Mileage/(weekdays out of service)
1934	5,169 (2)
1935	63,721 (75)
1936	53,889 (93)
1937	42,524 (66)
1938	41,004 (79)
1939	42,808 (60)
1940	39,714 (42)
1941	43,071 (24)
1942	40,927 (49)
1943	52,878 (52)
1944	44,693 (75)
1945	36,317 (81)
1946	43,782 (67)
1947	30,952 (112)
1948	29,266 (73)
1949	35,897 (49)
1950	29,106 (51)
1951	38,331 (54)
1952	43,733 (35)
1953	33,699 (52)
1954	35,024 (69)
1955	37,046 (75)
1956	41,780 (55)
1957	44,339 (31)
1958	37,564 (55)
1959	36,525
1960	38,450

Mileage at 12/36: 122,779
Mileage at 31/12/50: 675,718

Sheds
Crewe	17/11/34
Kentish Town	10/2/35
Llandudno Jct	14/11/36
Rugby	5/6/43
Crewe South	8/5/48
Aston	5/6/48
Monument Lane	20/9/58
Aston	2/7/60
Saltley	16/1/65
Aston	6/3/65
Tyseley	2/10/65
Shrewsbury	12/11/66
Chester	11/3/67
Stoke	15/4/67
Crewe South	12/8/67

Withdrawn w.e. 30/9/67

Rugby's 5052 at Birmingham New Street between October 1945 (when it was fitted with a domed boiler) and May 1948 when it was transferred to Crewe South. Photograph J.T. Clewley, www.transporttreasury.co.uk

45052 at an unknown date, probably late 1940s, with a comparatively rare scroll pattern BR numberplate which is indicative of its early date of renumbering, w.e. 15/5/48. It was shedded at 3D Aston from 5/6/48 and carried domed boilers between 1945 and 1956. It was paired with riveted tenders throughout its life. Photograph E. Kearns.

45053

Built as 5053 at Vulcan Foundry 17/11/34
Renumbered 45053 w.e. 12/6/48

Improvements and modifications
28/3/38	Removal of vacuum pump
2/10/43	Steam sanding
?	Fitting BR ATC equipment

Repairs

20/11/35-20/12/35	LS
26/7/37-26/8/37	LS
3/3/38-28/3/38	HO
2/12/38-29/12/38	HS
13/10/39-3/11/39	LS
8/4/40-2/5/40	LO
1/11/40-23/11/40	LS
24/2/41-8/3/41	LO
1/9/41-8/10/41	HG
24/9/42-7/11/42	HS
16/6/43-26/7/43	HS
29/3/44-1/4/44	LO
24/4/44-24/5/44	HS
6/12/44-27/1/45	LO
23/6/45-11/8/45	LS
24/12/45-9/2/46	LS
1/6/46-29/6/46	LO
15/2/47-9/4/47	HG
8/11/47-22/11/47	LO
22/5/48-12/6/48	LS
25/6/49-29/8/49	LI
11/10/49-12/10/49	NC
19/12/49-25/1/50	LC
24/6/50-19/8/50	HI
2/12/50-29/12/50	LC
28/2/52-7/6/52	G
27/5/53-24/7/53	LI
14/1/54-16/1/54	NC(EO)
25/3/54-14/4/54	HI
17/1/55-12/2/55	LI
17/3/55-23/3/55	LC(EO)
16/5/55-19/5/55	NC(EO)
28/7/55-6/8/55	NC(EO)
28/9/55-15/10/55	LC(EO)
27/2/56-23/3/56	LI
6/4/56-21/4/56	NC(EO)
4/5/56-19/5/56	NC(EO)
1/6/56-23/6/56	NC(EO)
8/10/56-11/10/56	NC(EO)
24/1/57-22/2/57	G
10/4/57-18/4/57	LC
11/1/58-19/2/58	HI
12/5/58-22/5/58	LC(EO)
20/12/58-17/1/59	HI
13/11/59-18/12/59	HI
29/1/60-12/2/60	LC(EO)
4/4/60-14/4/60	LC(EO)
26/4/60-30/4/60	NC(EO)
14/9/60-29/9/60	LC(EO)
16/2/61-1/3/61	LC(EO)
20/3/61-13/5/61	G
15/5/61-1/6/61	LC
27/5/63-28/6/63	HI
9/7/63-10/7/63	NC
30/10/64-7/11/64	NC
3/3/65-20/3/65	LC

Boilers
New	8670
28/3/38	9019 from 5089
26/7/43	8929 from 5181
9/4/47	8965 from 5169

Tenders
New	9107
25/3/38	9041
22/2/41	9258 (welded)
18/6/43	9083
21/6/43	9068
23/5/44	9069
6/12/44	9060
6/2/45	9032
?	10594 (part-welded)
25/3/54	10550 (part-welded)
15/4/54	10686 (part-welded)
23/3/56	10619 (part-welded)
22/2/57	10546 (part-welded)
13/5/61	10719 (part-welded)
19/5/61	9833 (welded)

Mileage/(weekdays out of service)
1934	4,883 (3)
1935	46,160 (141)
1936	60,574 (74)
1937	46,503 (89)
1938	53,658 (77)
1939	58,021 (39)
1940	44,818 (67)
1941	40,442 (95)
1942	64,459 (65)
1943	57,338 (56)
1944	45,839 (72)
1945	48,285 (90)
1946	44,998 (88)
1947	45,132 (81)
1948	53,927 (56)
1949	39,841 (115)
1950	31,355 (128)

Mileage at 12/36: 111,617
Mileage at 31/12/50: 782,633

Sheds
Carlisle	17/11/34
Kentish Town	23/2/35
Trafford Park	6/7/35
Kentish Town	24/8/35
Carlisle N	28/11/36
Perth	17/4/37
Carlisle N	22/5/37
Inverness	30/5/42
Perth	22/2/53
Corkerhill	25/6/59
Ardrossan	10/10/60
Dalry Road	14/6/61
St Margarets	4/10/65

Withdrawn 11/11/66

45053 at Perth, its home depot, in 1954. It was domed between June 1952 and February 1957 and has a part-welded tender which it kept until 1961. Note the 'squeezed together' cab number to clear the tablet exchanger.

Sloping throatplate boilered 45054 with an express at Roade in 1952 still has LMS insignia on the welded tender it received in May of that year, although the locomotive had been renumbered in June 1948.

45054

Built as 5054 at Vulcan Foundry 17/11/34
Renumbered 45054 w.e. 5/6/48

Improvements and modifications

27/1/37	Sloping throatplate boiler
20/4/38	Removal of vacuum pump
8/8/39	BTH speed indicator
4/4/46	Steam sanding
12/4/57	Modification
13/8/59	Fitting BR ATC equipment

Repairs

Date	Type
23/10/35-10/1/36	HS
29/1/36-17/2/36	LO
6/1/37-11/2/37	HG
21/3/38-20/4/38	HS
21/6/39-8/8/39	LS
26/10/40-15/11/40	HG
2/1/43-23/1/43	LS
21/6/44-13/7/44	LS
27/11/44-16/12/44	LO
13/3/46-4/4/46	HG
19/7/47-1/9/47	HS
1/5/48-31/5/48	LO
23/2/49-28/3/49	LI
25/9/50-18/10/50	HG
6/5/52-29/5/52	LI
5/6/52-1/7/52	LC(EO)
9/11/53-15/12/53	HG
14/3/55-14/4/55	HI
20/9/55-14/10/55	LC(EO)
16/3/57-12/4/57	HG
4/7/59-13/8/59	HI
24/7/61-21/9/61	HG
8/5/62-26/5/62	LC
1/4/65-17/4/65	INT

Boilers

New	8671
27/1/37	New (sloping throatplate)
15/4/40	9457 from 5337 (sloping throatplate)
4/4/46	9536 from 5288 (sloping throatplate)
18/10/50	9570 from 5421 (sloping throatplate)
15/12/53	11940 from 45242 (sloping throatplate)
12/4/57	12134 from 45370 (sloping throatplate)
21/9/61	9536 from 45397 (sloping throatplate)

Tenders

Date	Tender
New	9108
6/5/52	10529 (welded)

Mileage/(weekdays out of service)

Year	Mileage (weekdays out of service)
1934	5,832 (-)
1935	57,424 (103)
1936	61,425 (75)
1937	46,015 (63)
1938	51,573 (60)
1939	38,901 (73)
1940	34,640 (65)
1941	37,462 (28)
1942	32,548 (20)
1943	34,331 (48)
1944	25,497 (78)
1945	30,675 (51)
1946	32,502 (52)
1947	35,322 (75)
1948	45,151 (53)
1949	43,856 (55)
1950	43,097 (45)
1951	48,462 (29)
1952	42,211 (67)
1953	39,485 (65)
1954	59,997 (38)
1955	42,293 (70)
1956	44,815 (33)
1957	45,397 (53)
1958	38,731 (62)
1959	41,006
1960	45,252

Mileage at 12/36: 124,681
Mileage at 31/12/50: 656,251

Sheds

Shed	Date
Carlisle	17/11/34
Kentish Town	23/2/35
Longsight	28/11/36
Carnforth	8/1/38
Preston	3/12/38
Chester	30/12/39
Mold Jct	25/5/40
Edge Hill	8/5/43
Crewe North	24/1/48
Barrow	14/2/48
Carnforth	15/9/51

Withdrawn w.e. 10/2/68

Below. A long way from its home shed at Carnforth, 45054 poses at Hereford in April 1957 when it was on a running-in turn from Crewe Works following Heavy General repair. It had sloping throatplate boilers since January 1937 and a welded tender since May 1952.

45055

Built as 5055 at Vulcan Foundry 24/11/34
Renumbered 45055 w.e. 15/1/49

Improvements and modifications
12/5/39	Removal of vacuum pump
10/1/41	BTH speed indicator
21/11/45	Steam sanding
1/2/58	Modification
7/5/60	Fitting BR ATC equipment

Repairs
30/9/35-29/10/35	LS
17/11/36-1/12/36	LS
21/4/37-10/5/37	LO
18/10/37-18/11/37	HG
27/4/39-12/5/39	LS
14/12/40-10/1/41	HG
9/2/42-7/3/42	LO
18/1/43-13/2/43	LS
2/5/44-16/5/44	LS
17/6/44-21/7/44	LO
10/10/45-21/11/45	HG
17/4/47-22/5/47	HS
25/11/48-12/1/49	HG
11/8/50-1/9/50	LI
5/12/51-18/1/52	LI
23/1/52-1/2/52	NC(Rect)
16/12/53-16/1/54	HG
27/1/56-25/2/56	HI
30/12/57-1/2/58	HG
26/4/60-7/5/60	NC(EO)
17/8/60-24/9/60	LI
4/2/63-7/3/63	HG
27/4/66-3/6/66	LI

Boilers
New	8672
4/11/37	8667 from 5050 (domed)
10/1/41	8917 from 5036
21/11/45	8954 from 5118
12/1/49	8996 from 5185
16/1/54	8905 from 45187
1/2/58	8660 from 45185 (domed)
?	8964

Tenders
New	9109
21/7/44	9515 (welded)

Mileage/(weekdays out of service)
1934	4,625 (3)
1935	70,372 (61)
1936	57,729 (111)
1937	37,190 (67)
1938	49,845 (36)
1939	40,694 (39)
1940	31,553 (51)
1941	39,109 (46)
1942	33,694 (38)
1943	36,996 (38)
1944	34,782 (69)
1945	34,287 (50)
1946	42,083 (15)
1947	39,070 (59)
1948	33,993 (66)
1949	46,709 (43)
1950	36,471 (56)
1952	38,336 (49)
1953	26,524 (42)
1954	30,750 (39)
1955	33,670 (63)
1956	37,479 (59)
1957	40,927 (40)
1958	37,139 (63)
1959	40,131
1960	36,684

Mileage at 12/36: 132,726
Mileage at 31/12/50: 669,202

Sheds
Kentish Town	24/11/34
Trafford Park	22/2/36
Tebay	14/11/36
Crewe North	3/7/37
Patricroft	14/8/37
Springs Branch	20/9/52
Mold Jct	7/5/55
Springs Branch	15/9/62
Southport	20/7/63
Bank Hall	18/6/66
Aintree	22/10/66
Warrington	13/5/67
Edge Hill	7/10/67
Patricroft	11/5/68
Lostock Hall	6/7/68

Stored
13/1/68-26/2/68

Withdrawn w.e. 3/8/68

Below. A typical early 1950s view of 45055 which was allocated to 10A Springs Branch from September 1952 to May 1955. It had been domeless since 1941 and had a welded tender from 1944. Photograph www.rail-online.co.uk

45056

Built as 5056 at Vulcan Foundry 24/11/34
Renumbered 45056 w.e. 23/10/48

Improvements and modifications

17/5/38	Removal of vacuum pump
14/6/40	BTH speed indicator
18/2/44	Steam sanding
27/2/59	Fitting BR ATC equipment

Repairs

8/10/35-13/11/35	LS
29/10/36-4/12/36	LS
8/4/37-26/4/37	LO
21/10/37-24/11/37	HG
28/3/38-17/5/38	LO
4/10/38-17/11/38	HS
24/4/39-23/5/39	LS
28/5/40-14/6/40	HG
20/12/41-23/1/42	LS
17/7/42-6/8/42	LO
6/1/43-27/1/43	HS
1/2/44-19/2/44	HG
10/4/44-2/5/44	LO
2/3/45-23/3/45	LO
11/12/45-14/1/46	LS
24/3/47-14/5/47	HS
7/11/47-4/12/47	HS
21/9/48-20/10/48	HG
14/12/49-10/1/50	LI
27/8/51-28/9/51	LI
5/5/52-28/5/52	LC(EO)
4/6/53-6/7/53	HG
21/1/54-20/2/54	LC(EO)
17/2/55-16/3/55	HI
15/8/55-17/9/55	LC(EO)
11/2/57-20/3/57	HG
18/9/57-9/10/57	LC(EO)
1/10/58-30/10/58	LI
19/2/59-27/2/59	NC(EO)
18/12/59-4/2/60	LI
24/5/61-19/6/61	HI
11/6/63-26/7/63	HG
26/9/66-15/11/66	NC

Boilers

New	8673
10/11/37	8642 from 5025 (domed)
4/6/40	8999 from 5219
19/2/44	9059 from 5070
20/10/48	8926 from 5071
6/7/53	9012 from 45130
20/3/57	8833 from 45140
?	8980

Tenders

New	9110
7/4/38	9292 (welded)
6/5/44	9500 (welded)
14/1/46	9325 (welded)
20/3/57	9238 (welded)

Mileage/(weekdays out of service)

Year	Mileage (weekdays out of service)
1934	4,537 (-)
1935	66,225 (73)
1936	44,242 (113)
1937	36,923 (139)
1938	67,033 (112)
1939	57,922 (71)
1940	39,004 (52)
1941	30,616 (59)
1942	31,286 (74)
1943	46,643 (38)
1944	37,897 (64)
1945	33,919 (75)
1946	43,332 (38)
1947	29,124 (83)
1948	50,378 (48)
1949	45,341 (55)
1950	37,525 (39)
1951	32,384 (60)
1952	40,792 (48)
1953	36,712 (67)
1954	40,615 (77)
1955	40,600 (110)
1956	41,474 (92)
1957	42,362 (92)
1958	40,266 (62)
1959	41,810
1960	42,567

Mileage at 12/36: 115,004
Mileage at 31/12/50: 701,427

Sheds

Kentish Town	24/11/34
Nottingham	7/3/36
Kentish Town	25/9/37
Toton	18/5/40
Saltley	4/7/42
Bath	16/12/44
Sheffield	1/10/49
Millhouses	20/3/54
Holyhead	26/4/58
Llandudno Jct	11/3/61
Rugby	23/9/61
Nuneaton	7/3/64 (loan)
Rugby	18/7/64
Northampton	9/1/65
Rugby	27/2/65
Crewe South	10/4/65
Speke Junction	18/3/67 (loan)
Speke Junction	1/4/67

Withdrawn w.e. 19/8/67

Midland Division 5056 was allocated to 14B Kentish Town from new in November 1934 until March 1936 when it went to Nottingham. It still has a crosshead vacuum pump which was removed in May 1938 and also prominent top feed pipes which would have been recessed when it received a domeless boiler during a works visit in late 1937.

Holyhead's 45056 is blowing-off furiously as it backs onto its train at Bangor on 23 June 1959. It has AWS, fitted four months earlier and has acquired one of the ten sets of original Vulcan Foundry scalloped steam pipe casings. 45056 had domeless boilers from June 1940 until withdrawn in August 1967 and welded tenders from April 1938 onwards. Photograph Alec Swain, www.transporttreasury.co.uk

45057

Built as 5057 at Vulcan Foundry 1/12/34
Renumbered 45057 w.e. 8/5/48

Improvements and modifications
23/12/37	Sloping throatplate boiler
20/6/39	Removal of vacuum pump
4/6/40	BTH speed indicator
4/11/44	Steam sanding
14/5/59	Fitting BR ATC equipment

Repairs
21/10/35-27/11/35	LS
23/7/36-18/8/36	LS
10/10/36-16/11/36	LO
3/3/37-1/4/37	LO
28/5/37-14/6/37	LO
27/11/37-23/12/37	HG
12/5/39-20/6/39	LS
15/5/40-4/6/40	HG
29/10/41-15/11/41	LS
23/1/42-13/2/43	LO
22/7/43-9/8/43	LS
21/2/44-11/3/44	LO
14/9/44-4/11/44	HG
11/1/46-9/2/46	HS
24/5/47-26/6/47	HS
6/4/48-5/5/48	LO
18/10/48-17/11/48	HG
22/9/50-28/10/50	HI
29/11/51-2/1/52	HI
30/6/53-17/8/53	HG
28/4/55-8/6/55	HI
28/4/56-30/5/56	LC
1/3/58-30/5/58	HG
29/4/59-14/5/59	NC(EO)
25/5/59-12/6/59	LC
15/11/60-16/12/60	HI
18/10/62-10/11/62	HI
11/9/64-10/10/64	LI
24/5/66-1/7/66	LC

Boilers
New	8674
8/12/37	10129 from New (sloping throatplate)
4/6/40	9735 from 5040 (sloping throatplate)
4/11/44	9395 from 5260 (sloping throatplate)
17/11/48	11917 from 4818 (sloping throatplate)
17/8/53	9353 from 44848 (sloping throatplate)
30/5/58	9454 from 45387 (sloping throatplate)

Tenders
New	9111
@31/12/39	9113
21/7/43	9687 (welded)
9/2/46	9531 (welded)

Mileage/(weekdays out of service)
1934	4,449 (3)
1935	58,515 (94)
1936	59,307 (94)
1937	41,013 (94)
1938	51,007 (59)
1939	45,010 (64)
1940	43,161 (47)
1941	37,749 (59)
1942	30,509 (33)
1943	32,881 (40)
1944	33,268 (89)
1945	50,608 (57)
1946	43,306 (75)
1947	35,497 (87)
1948	34,092 (80)
1949	38,853 (36)
1950	33,179 (50)
1951	34,096 (60)
1952	42,984 (55)
1953	34,778 (75)
1954	36,739 (30)
1955	35,009 (53)
1956	35,302 (57)
1957	31,927 (30)
1958	23,891 (93)
1959	31,086
1960	33,964

Mileage at 12/36: 122,271
Mileage at 31/12/50: 672,404

Sheds
Kentish Town	1/12/34
Carlisle N	5/12/36 (on loan)
Carlisle W	19/12/36
Crewe North	4/9/37
Carlisle W	25/9/37
Carnforth	19/7/41
Preston	21/3/42
Patricroft	4/4/42
Edge Hill	9/10/43
Crewe North	13/5/44
Rugby	10/6/44
Bletchley	8/5/48
Patricroft	7/7/51
Chester	31/12/55
Springs Branch	25/8/56
Newton Heath	27/7/63
Speke Junction	5/10/63

Withdrawn w.e. 12/8/67

Springs Branch allocated 45057 with a long train of mineral wagons, including four wooden ex-Private Owners at the front, passes over the water troughs at Halebank on Merseyside on 2 March 1959. It was one of the pre-war conversions to sloping throatplate boiler, in December 1937, and got its welded tender in 1943.

45057 in store at Speke Junction following its withdrawal in August 1967. It remained there until June 1968 when it went for scrapping to Cohens at Kettering. 45057 had been fitted with AWS in May 1959 and had a sloping throatplate boiler from 1937 onwards.

45058

Built as 5058 at Vulcan Foundry 1/12/34
Renumbered 45058 w.e. 11/6/49

Improvements and modifications

16/11/37	Sloping throatplate boiler
22/4/39	Removal of vacuum pump
18/7/45	Steam sanding
20/1/59	Fitting BR ATC equipment

Repairs

19/10/35-18/11/35	LS
11/9/36-14/10/36	LS
1/3/37-12/4/37	LO
3/11/37-30/11/37	HG
13/3/39-3/5/39	LS
17/1/41-4/2/41	HG
11/2/43-1/3/43	LS
21/4/44-5/5/44	LS
25/7/44-5/8/44	LO
13/6/45-18/7/45	HG
12/1/47-6/2/47	HS
17/2/48-10/3/48	LS
14/5/49-9/6/49	LI
5/9/50-27/9/50	HG
23/2/52-28/3/52	LI
22/8/53-10/9/53	HI
13/4/55-27/5/55	HG
22/9/56-18/10/56	LI
18/12/57-2/1/58	LC (EO)
8/8/58-29/8/58	LI
12/1/59-20/1/59	NC (EO)
9/2/60-11/3/60	HG
29/1/62-28/2/62	LI
29/4/64-23/5/64	LI
10/6/64-10/6/64	NC
18/6/64-19/6/64	LC
18/2/65-20/2/65	LC

Boilers

New	8675
16/11/37	9742 New (sloping throatplate)
4/2/41	9486 from 5366 (sloping throatplate)
18/7/45	9349 from 5252 (sloping throatplate)
27/9/50	9416 from 5245 (sloping throatplate)
27/5/55	9379 from 45243 (sloping throatplate)
11/3/60	9465 from 45454 (sloping throatplate)

Tenders

New	9112

Mileage/(weekdays out of service)

Year	Mileage (weekdays out of service)
1934	3,453 (1)
1935	59,224 (74)
1936	63,253 (80)
1937	43,131 (90)
1938	65,890 (33)
1939	47,637 (70)
1940	39,931 (50)
1941	33,075 (49)
1942	30,603 (38)
1943	33,193 (49)
1944	37,384 (65)
1945	32,697 (66)
1946	33,607 (42)
1947	37,286 (42)
1948	35,633 (49)
1949	35,721 (45)
1950	33,234 (43)
1951	42,105 (30)
1952	38,300 (55)
1953	37,801 (41)
1954	39,867 (36)
1955	38,046 (82)
1956	32,884 (73)
1957	40,346 (52)
1958	35,901 (51)
1959	37,523
1960	38,315

Mileage at 12/36: 125,930
Mileage at 31/12/50: 664,952

Sheds

Kentish Town	1/12/34
Carlisle N	5/12/36
Carlisle W	19/12/36
Crewe	4/9/37
Carlisle W	25/9/37
Carnforth	19/7/41
Willesden	4/10/41
Edge Hill	8/11/41
Carlisle W	7/8/43
Aston	14/7/45
Saltley	16/1/65
Aston	6/3/65
Shrewsbury	16/10/65

Withdrawn w.e. 8/10/66

Below. 45058, shedded at Aston since 1945, pictured at Crewe on 25 June 1960 was a pre-war sloping throatplate conversion, in November 1937. It has kept its original riveted tender. Visible recent modifications are the external steam lance pipe and AWS fitted January 1959. Photograph D. Forsyth, Paul Chancellor Collection.

45059

Built as 5059 at Vulcan Foundry 8/12/34
Renumbered 45059 w.e. 22/6/49

Improvements and modifications
17/5/39	Removal of vacuum pump
20/12/41	Steam sanding
14/7/45	Sloping throatplate boiler
9/5/58	Modification
25/1/64	Fitting Smith-Stone speedometer

Repairs
24/3/36-7/4/36	LS
11/1/37-23/1/37	LO
11/3/37-15/4/37	HG
25/3/39-17/5/39	LS
1/6/40-15/6/40	LO
15/9/41-22/9/41	LO
4/12/41-20/12/41	HG
6/11/42-3/12/42	HS
29/8/44-16/9/44	LS
26/5/45-14/7/45	HG
21/8/47-23/9/47	LS
27/1/48-5/3/48	LO
16/12/48-22/1/49	HI
9/3/50-6/4/50	HG
18/8/51-26/9/51	LI
15/1/53-19/2/53	LI
27/10/53-26/11/53	LC(EO)
5/5/54-4/6/54	HG
21/1/56-18/2/56	HI
22/2/57-22/3/57	LI
12/2/58-9/5/58	HG
15/2/60-18/3/60	LI
8/10/63-14/11/63	HG

Boilers
New	8676
1/4/37	8644 from 5027 (domed)
20/12/41	8930 from 5216
14/7/45	9356 from 5323 (sloping throatplate)
6/4/50	9486 from 5386 (sloping throatplate)
4/6/54	11336 from 45338 (sloping throatplate)
9/5/58	11918 from 45280 (sloping throatplate)
?	11909 (sloping throatplate)

Tenders
New	9113
@31/12/39	9305 (welded)
3/12/42	9315 (welded)
14/6/45	9472 (welded)
14/11/63	9268 (welded)

Mileage/(weekdays out of service)
1934	4,089 (-)
1935	70,509 (12)
1936	38,462 (34)
1937	35,373 (65)
1938	34,469 (38)
1939	28,118 (66)
1940	28,690 (83)
1941	29,654 (59)
1942	35,887 (50)
1943	38,101 (38)
1944	29,906 (44)
1945	30,844 (68)
1946	36,054 (42)
1947	26,485 (81)
1948	32,591 (81)
1949	34,520 (68)
1950	37,385 (50)
1951	29,360 (79)
1952	41,891 (34)
1953	35,847 (99)
1954	34,509 (64)
1955	27,001 (81)
1956	49,432 (41)
1957	43,890 (69)
1958	27,874 (146)
1959	35,195
1960	32,104

Mileage at 12/36: 113,060
Mileage at 31/12/50: 571,137

Sheds
Crewe	8/12/34
Carlisle	10/2/35
Crewe	19/10/35
Patricroft	16/11/35
Crewe North	27/3/43
Nottingham	25/10/47 (loan)
Nottingham	6/12/47
Saltley	20/10/51
Sheffield	10/10/53
Cricklewood	20/11/54
Derby	10/2/62
Leicester	26/5/62 (loan)
Derby	23/6/62
Burton	22/9/62
Speke Jct	13/3/65 (loan)
Speke Jct	3/4/65

Withdrawn w.e. 8/7/67

Below. A broadside view of 45059 in the mid-1960s at Shrewsbury showing the Smith-Stone speed indicator driven off the rear coupled axle which was fitted as late as January 1964, though interestingly there is no AWS. It was one of the later conversions to sloping throatplate boiler in July 1945 and had a welded tender from before the war. Photograph www.rail-online.co.uk

45060

Mileage at 12/36: 90,169
Mileage at 31/12/50: 550,904

Built as 5060 at Vulcan Foundry 15/12/34
Renumbered 45060 w.e. 28/8/48

Improvements and modifications
16/9/38	Removal of vacuum pump
?	Steam sanding
17/11/56	Modification
10/4/59	Fitting BR ATC equipment

Repairs
20/1/36-4/2/36	LS
12/11/36-19/12/36	LO
23/3/37-8/4/37	HS
20/8/38-16/9/38	HG
12/6/40-29/6/40	LO
14/1/41-29/1/41	HS
16/9/41-18/10/41	LO
25/1/43-19/2/43	HG
25/5/44-9/6/44	LS
7/7/45-4/8/45	LO
24/11/45-27/12/45	LS
30/1/47-8/3/47	HG
6/8/48-28/8/48	HS
23/10/50-17/11/50	LI
25/4/51-31/5/51	LC
13/8/52-18/9/52	HG
28/9/53-16/10/53	HI
3/4/55-28/4/55	LI
26/9/56-17/11/56	HG
10/3/59-10/4/59	LI
23/1/61-24/2/61	HI

Boilers
New	8677
16/9/38	9053 from 5123
19/2/43	8906 from 5207
8/3/47	8913 from 5149
18/9/52	8677 from 45209 (domed)
17/11/56	8639 from 45039 (domed)

Tenders
New	9114
27/12/45	9082
26/3/52	9092

Mileage/(weekdays out of service)
1934	2,419 (-)
1935	49,993 (58)
1936	37,757 (91)
1937	43,619 (71)
1938	30,860 (116)
1939	43,444 (39)
1940	25,510 (77)
1941	32,367 (73)
1942	30,464 (43)
1943	37,193 (45)
1944	30,380 (56)
1945	25,973 (87)
1946	34,417 (49)
1947	35,356 (58)
1948	27,942 (72)
1949	25,482 (50)
1950	37,778 (54)
1951	36,336 (68)
1952	34,975 (51)
1953	40,426 (47)
1954	43,316 (33)
1955	36,066 (63)
1956	33,631 (69)
1957	45,623 (21)
1958	42,050 (32)
1959	38,652
1960	33,783

Sheds
Farnley Jct	29/12/34
Wakefield	@4/2/36
Huddersfield	24/4/37
Wakefield	9/11/40
Sowerby Bridge	2/8/41
Wakefield	7/2/42
Crewe South	6/3/43 (loan)
Crewe South	10/4/43
Stoke	20/6/59
Bescot	3/9/60
Stoke	10/9/60

Stored
17/10/38-22/12/38

Withdrawn w.e. 4/3/67

Below. 45060 was a Central Division engine until 1942 and was allocated to Stoke in 1960 where it was pictured on 7 October 1964. It was domed from September 1952 until withdrawn and the blanking plates in the cladding on the firebox shoulders clearly show the that the current boiler was one of those rebuilt with additional superheating in the late-1930s. 45060 was on its third riveted tender and acquired its AWS in April 1959. Photograph www.rail-online.co.uk

45061

Built as 5061 at Vulcan Foundry 15/12/34
Renumbered 45061 w.e. 1/5/48

Improvements and modifications

5/12/38	Steam sanding
5/12/38	Removal of vacuum pump
13/11/40	BTH speed indicator
23/12/54	Modification
15/12/61	Fitting BR ATC equipment

Repairs

24/2/36-9/3/36	LS
9/6/36-11/6/36	LO
24/6/36-27/6/36	LO
6/7/37-29/7/37	HS
4/10/37-3/11/37	LO
9/11/38-5/12/38	HG
26/10/40-13/11/40	LS
23/3/42-21/4/42	HS
22/2/43-15/3/43	HG
5/1/44-12/2/44	LS
24/4/45-12/5/45	LS
5/11/46-18/12/46	HS
25/3/48-28/4/48	HG
13/9/48-3/11/48	NC
25/1/50-14/2/50	HG
24/2/50-2/3/50	NC Rect
14/3/50-17/3/50	Painting Only
18/12/51-12/1/52	LI
19/1/52-23/1/52	Rect EO
5/12/52-7/1/53	HI
28/7/53-8/8/53	LC(EO)
1/11/54-23/12/54	HG
15/4/57-10/5/57	HI
15/1/58-15/2/58	LC(EO)
18/2/58-11/3/58	NC(EO)
14/9/59-16/10/59	HG
5/12/61-15/12/61	NC
18/9/62-12/10/62	HI
15/3/65-10/4/65	LI

Boilers

New	8678
5/12/38	8918 from 5138
15/3/43	9036 from 5000
28/4/48	8643 from 5191 (domed)
14/2/50	9020 from 5107
23/12/54	8973 from 45065
16/10/59	8943 from 45065

Tenders

New	9115
5/1/44	9230 (welded)

Mileage/(weekdays out of service)

1934	2,059 (-)
1935	44,535 (82)
1936	37,897 (97)
1937	34,335 (93)
1938	39,698 (71)
1939	47,763 (39)
1940	29,283 (84)
1941	36,040 (40)
1942	34,722 (58)
1943	45,136 (38)
1944	33,212 (69)
1945	32,861 (69)
1946	30,501 (76)
1947	32,485 (60)
1948	27,993 (114)
1949	33,191 (60)
1950	32,069 (64)
1951	34,715 (54)
1952	44,960 (82)
1953	45,439 (39)
1954	30,883 (95)
1955	39,653 (51)
1956	38,328 (40)
1957	38,809 (54)
1958	35,618 (91)
1959	36,047
1960	37,309

Mileage at 12/36: 84,491
Mileage at 31/12/50: 573,780

Sheds

Farnley Jct	29/12/34
Wakefield	@11/6/36
Huddersfield	24/4/37
Southport	11/10/47
Sheffield	8/7/50 (loan)
Southport	5/8/50
Blackpool	10/3/51 (loan)
Southport	28/4/51
Blackpool	9/6/51 (loan)
Blackpool	30/6/51
Southport	27/10/51 (loan)
Accrington	19/1/52
Rose Grove	6/9/52
Southern Region	23/5/53 (loan)
Rose Grove	20/6/53
Blackpool	4/7/53
Southport	25/9/54
Aintree	17/3/62
Carlisle Kingmoor	18/5/63

Withdrawn w.e. 18/11/67

45061 in the mid-1960s with, appropriately, a Class 5 fitted freight. It was one of those sent briefly to the SR in May/June 1953 and was domeless except between April 1948 and February 1950. It had a welded tender from January 1944 and was fitted with AWS in December 1961.

Cricklewood 's 45062 on 17 May 1960 at Cheadle Heath on the ex-MR main line from Manchester to London. It only carried one domed boiler, from October 1950 to April 1956, though it alternated between riveted and welded tenders. Photograph D. Forsyth, Paul Chancellor Collection

45062

Built as 5062 at Vulcan Foundry 15/12/34
Renumbered 45062 w.e. 4/6/49

Improvements and modifications

10/10/38	Removal of vacuum pump
?	Steam sanding
17/4/56	Modification
1/3/63	Fitting BR ATC equipment

Repairs

15/1/36-3/2/36	LS
13/10/36-19/10/36	LO
9/3/37-2/4/37	HS
14/7/37-16/7/37	LO
6/9/37-16/9/37	LO
24/8/38-10/10/38	HG
1/10/39-3/10/39	TRO
12/12/39-28/12/39	HS
28/1/41-13/3/41	LS
10/1/42-31/1/42	HG
2/3/43-30/3/43	LS
26/9/44-7/10/44	LS
22/3/46-12/4/46	HG
18/12/47-24/1/48	LS
13/5/49-2/6/49	LI
28/8/50-13/10/50	HG
28/4/52-19/5/52	LI
18/4/54-11/5/54	HI
18/12/54-12/1/55	LC
16/3/56-17/4/56	HG
28/1/58-13/3/58	HI
25/8/58-26/9/58	LC(EO)
1/6/59-16/6/59	LC(EO)
2/1/61-14/3/61	HI
1/2/63-1/3/63	HG

Boilers

New	8679
10/10/38	9058 from 5072
31/1/42	8833 from 5172
12/4/46	8928 from 5139
13/10/50	8646 from 5110 (domed)
17/4/56	9034 from 45001
?	8918

Tenders

New	9116
3/2/36	9230 (welded)
5/1/44	9115
15/10/54	10569 (welded)
26/1/55	9115

Mileage/(weekdays out of service)

1934	1,378 (-)
1935	51,113 (49)
1936	43,909 (87)
1937	30,676 (100)
1938	43,279 (70)
1939	49,441 (92)
1940	38,581 (33)
1941	28,505 (80)
1942	39,836 (40)
1943	41,248 (51)
1944	36,189 (53)
1945	40,209 (41)
1946	40,771 (51)
1947	32,178 (57)
1948	39,488 (81)
1949	45,566 (48)
1950	30,641 (95)
1951	38,640 (43)
1952	36,230 (49)
1953	37,952 (68)
1954	31,919 (79)
1955	36,793 (72)
1956	38,365 (69)
1957	37,809 (61)
1958	40,064 (100)
1959	39,904
1960	22,839

Mileage at 12/36: 96,400
Mileage at 31/12/50: 633,006

Sheds

Farnley Jct	29/12/34
Newton Heath	23/2/35
Farnley Jct	19/8/35
Huddersfield	6/11/37
Low Moor	16/7/38
Accrington	12/7/41
Huddersfield	11/7/42
Low Moor	14/11/42
Bank Hall	15/5/48
Farnley Jct	29/4/50
Sheffield	11/11/50 (loan)
Sheffield	2/2/52
Derby	21/9/57
Cricklewood	10/1/59
Derby	17/2/62
Burton	21/9/62
Derby	23/11/63
Burton	12/9/64
Agecroft	3/7/65 (loan)
Agecroft	24/7/65
Trafford Park	22/10/66

Stored
2/10/66-29/10/66

Withdrawn w.e. 15/4/67

45063

Built as 5063 at Vulcan Foundry 22/12/34
Renumbered 45063 w.e. 19/6/48

Improvements and modifications
21/9/39	Steam sanding
?	Removal of vacuum pump
26/5/55	Modification
17/1/61	Fitting BR ATC equipment

Repairs
27/1/36-10/2/36	LS
7/7/36-10/7/36	LO
9/3/37-1/4/37	HS
7/9/37-10/9/37	LO
4/5/38-3/6/38	LS
25/8/39-21/9/39	HG
2/12/40-19/12/40	LS
16/11/42-24/12/42	LS
23/7/43-7/8/43	LO
15/6/44-30/6/44	HG
26/11/45-29/12/45	LS
18/10/46-26/11/46	HS
20/5/48-17/6/48	LS
13/12/49-20/1/50	HG
1/10/51-1/11/51	LI
22/9/53-21/10/53	LI
25/4/55-26/5/55	HG
24/10/56-10/11/56	HI
12/11/56-22/11/56	NC(Rect)(EO)
25/11/57-1/1/58	LI
7/8/59-18/9/59	HG
6/1/61-17/1/61	NC(EO)
28/11/62-1/1/63	HI
16/9/65-21/10/65	INT

Boilers
New	8680
25/8/39	8664 from 5063 (domed)
30/6/44	8836 from 5203
20/1/50	8640 from 5032 (domed)
26/5/55	8663 from 45103 (domed)
18/9/59	8825 from 45070

Tenders
New	9117
24/7/43	9632 (welded)
29/12/45	9241 (welded)

Mileage/(weekdays out of service)
1934	751
1935	51,835 (54)
1936	39,034 (92)
1937	34,468 (85)
1938	36,498 (82)
1939	30,475 (94)
1940	30,295 (61)
1941	36,748 (39)
1942	30,757 (85)
1943	38,631 (61)
1944	42,412 (58)
1945	39,572 (87)
1946	28,001 (86)
1947	35,599 (62)
1948	34,907 (54)
1949	29,344 (48)
1950	41,397 (53)
1951	31,308 (71)
1952	35,958 (44)
1953	28,236 (63)
1954	36,150 (60)
1955	35,438 (76)
1956	40,132 (76)
1957	40,113

Mileage at 12/36: 91,620
Mileage at 31/12/50: 580,724

Sheds
Farnley Jct	29/12/34
Newton Heath	23/2/35
Farnley Jct	17/8/35
Blackpool	12/7/41
Farnley Jct	12/10/46
Newton Heath	11/7/53 (loan)
Farnley Jct	15/8/53
Neville Hill	1/3/64
Holbeck	7/6/64

Stored
7/10/36-8/11/36

Withdrawn 10/10/66

45063 had four spells at 55C Farnley Junction with the final one lasting from 1953 until March 1964 when it moved to Holbeck. It had domeless boilers between January 1950 and September 1959, and was fitted with AWS in June 1961 which therefore dates this photograph as early-1960s. Photograph G.W. Sharpe.

45064

Built as 5064 at Vulcan Foundry 22/12/34
Renumbered 45064 w.e. 1/4/50

Improvements and modifications
17/6/38	Removal of vacuum pump
23/3/43	Steam sanding
12/2/59	Fitting BR ATC equipment

Repairs
7/12/35-3/1/36	LS
21/5/36-17/7/36	LO
14/10/36-12/11/36	LS
3/3/37-30/3/37	LO
14/4/38-17/6/38	HG
1/11/39-23/11/39	LS
25/5/40-5/7/40	TRO
25/2/41-19/3/41	LS
24/12/41-9/1/42	LO
22/2/43-23/3/43	HG
21/9/44-7/10/44	LS
17/9/46-10/10/46	LS
21/2/48-25/3/48	HG
6/1/50-25/1/50	TRO
7/3/50-27/3/50	HI
12/1/52-23/2/52	HG
10/12/53-13/12/53	LC(EO)
19/3/54-12/4/54	HI
24/5/54-9/6/54	NC(EO)
29/8/55-12/10/55	HI
26/10/56-6/11/56	LC(EO)
25/5/57-26/6/57	HG
24/11/58-20/12/58	LI
5/2/59-12/2/59	NC(EO)
25/8/59-2/10/59	LC(EO)
10/4/61-9/5/61	LI
5/9/62-22/9/62	LC
10/4/64-6/6/64	HI
3/7/64	Rect

Boilers
New	8681
1/6/38	9050 from 5120
23/3/43	9015 from 5168
25/3/48	8673 from 5064 (domed)
23/2/52	8827 from 45216
26/6/57	8680 from 45134 (domed)

Tenders
New	9118
26/7/66	9223

Mileage/(weekdays out of service)
1934	354 (-)
1935	50,630 (117)
1936	38,858 (139)
1937	51,125 (71)
1938	44,866 (91)
1939	49,351 (75)
1940	35,667 (69)
1941	26,370 (51)
1942	37,762 (57)
1943	30,934 (51)
1944	30,717 (51)
1945	33,006 (40)
1946	28,717 (71)
1947	29,523 (54)
1948	32,539 (70)
1949	31,543 (101)
1950	36,526 (64)
1951	39,599 (44)
1952	45,586 (63)
1953	48,047 (28)
1954	39,709 (62)
1955	36,851 (87)
1956	38,754 (45)
1957	42,852 (55)
1958	35,750 (86)
1959	38,984
1960	34,389

Mileage at 12/36: 89,842
Mileage at 31/12/50: 588,288

Sheds
Newton Heath	29/12/34
Wakefield	26/1/35
Derby	6/4/35
Crewe North	28/11/36
Shrewsbury	25/9/37
Chester	7/5/38
Edge Hill	26/10/40
Crewe South	14/6/41
Crewe North	26/6/43
Crewe South	7/8/43
Willesden	5/6/48
Bescot	7/5/60
Northampton	3/11/62
Bletchley	26/1/63
Bescot	14/3/64
Stourbridge	26/3/66
Llandudno Jct	28/5/66
Chester	8/10/66

Stored
21/1/66-28/5/66

Withdrawn w.e. 11/3/67

45064 on a train of loaded minerals with six wooden five plank wagons leading. The period is some time after June 1959, when it received AWS, and November 1962, when it was transferred from Bescot to Northampton. 45064 had its second domed boiler and kept its original riveted tender throughout. Photograph www.transporttreasury.co.uk

45065 with assistance up Shap on a ten coach train in the early 1960s. It got AWS in February 1959 and was at Aston from November of that year until moving to Rugby in February 1963. Apart from two years in the 1930s, 45065 was always domeless; it ran with a welded tender from 1945 onwards. Photograph www.rail-online.co.uk

45065

Built as 5065 at Vulcan Foundry 29/12/34
Renumbered 45065 w.e. 2/10/48

Improvements and modifications
28/2/38	BTH speed indicator
28/2/38	Removal of vacuum pump
12/4//39	Steam sanding
24/2/59	Fitting BR ATC equipment

Repairs
15/1/36-24/2/36	LS
5/10/36-26/10/36	LO
10/2/37-1/4/37	HG
19/4/37-3/5/37	LO
19/1/38-28/2/38	HS
22/6/38-10/8/38	LO
6/3/39-12/4/39	HG
15/3/40-26/4/40	LS
29/4/41-31/5/41	HS
4/8/41-22/8/41	LO
27/3/42-5/5/42	LS
15/3/43-6/4/43	LS
18/1/44-18/2/44	HS
8/2/45-28/2/45	HG
2/10/45-1/11/45	LO
21/8/46-19/9/46	LS
6/12/46-19/12/46	LO
8/10/47-12/11/47	LS
10/6/48-15/7/48	LO
18/8/48-27/9/48	LO
8/6/49-23/6/49	HG
3/1/51-26/1/51	HI
29/5/52-21/6/52	LI
1/2/54-3/3/54	HG
17/12/55-11/1/56	HI
22/12/56-18/1/57	LI
23/12/57-15/1/58	LC(EO)
20/9/58-21/10/58	HG
17/2/59-24/2/59	NC(EO)
9/5/60-10/6/60	LI
29/6/60-19/8/60	LC(EO)
16/7/62-9/8/62	HI
19/6/64-4/7/64	HI
23/7/65-28/8/65	NC

Boilers
New	8682
15/3/37	8664 from 5047 (domed)
12/4/39	8907 from 5127
31/5/41	9021 from 5207
28/2/45	9027 from 5080
23/6/49	8973 from 5132
3/3/54	8943 from 45077
21/10/58	8905 from 45055

Tenders
New	9119
1/11/45	9564 (welded)
22/4/64	9660 (welded)

Mileage/(weekdays out of service)
1934	185 (-)
1935	60,898 (82)
1936	47,317 (123)
1937	61,824 (100)
1938	49,419 (148)
1939	52,385 (90)
1940	44,188 (71)
1941	40,612 (109)
1942	46,857 (80)
1943	47,314 (63)
1944	46,948 (59)
1945	43,014 (77)
1946	45,457 (76)
1947	44,060 (69)
1948	34,561 (111)
1949	42,850 (44)
1950	41,746 (41)
1951	40,838 (44)
1952	39,805 (41)
1953	37,684 (35)
1954	38,925 (58)
1955	34,591 (59)
1956	39,680 (43)
1957	43,711 (49)
1958	30,532 (78)
1959	43,518
1960	32,967

Mileage at 12/36: 108,400
Mileage at 31/12/50: 749,635

Sheds
Newton Heath	29/12/34
Kentish Town	6/4/35
Leeds	17/11/45
Crewe North	2/10/48 (loan)
Crewe North	6/11/48
Carlisle Upperby	1/10/49
Aston	15/9/51
Monument Lane	12/9/59
Aston	7/11/59
Rugby	23/2/63
Northampton	30/3/63
Willesden	22/6/63
Rugby	17/8/63
Crewe South	19/9/64
Rugby	28/11/64
Nuneaton	29/5/65
Heaton Mersey	28/5/66

Stored
1/1/66-27/4/66
8/6/67-27/11/67

Withdrawn w.e. 27/4/68

45066

Built as 5066 at Vulcan Foundry 5/1/35
Renumbered 45066 w.e. 19/6/48

Improvements and modifications

?	Removal of vacuum pump
20/5/45	Steam sanding
4/6/60	Sloping throatplate boiler
4/6/60	Fitting BR ATC equipment

Repairs

13/11/35-18/12/35	LS
14/7/36-5/8/36	LO
24/2/37-17/4/37	LS
16/3/38-31/3/38	LO
6/10/38-24/10/38	HS
8/4/39-4/5/39	HS
9/1/40-21/2/40	HG
27/1/41-20/2/41	LS
14/11/41-13/12/41	LS
18/12/41-2/1/42	LO
3/11/42-11/12/42	LS
20/9/43-13/10/43	HG
30/11/43-25/12/43	LO
12/8/44-6/9/44	LS
24/8/45-18/10/45	LS
28/10/46-11/1/47	HG
4/9/47-18/9/47	LO
8/5/48-19/6/48	LS
25/10/48-9/11/48	LO
10/9/49-28/10/49	LI
21/7/50-20/9/50	LC
16/6/51-20/10/51	G
13/12/52-23/1/53	LI
5/2/53-13/2/53	LC
30/3/53-11/4/53	LC
19/10/53-30/10/53	LC(EO)
7/4/54-21/5/54	LI
20/11/54-17/12/54	LC(EO)
17/9/55-14/10/55	G
29/12/56-31/1/57	HI
10/4/57-12/4/57	NC
1/2/58-10/3/58	LI
6/5/58-9/5/58	NC EO
27/1/59-3/3/59	LI
13/4/60-4/6/60	G
2/10/61-25/11/61	HI
6/8/63-13/9/63	LI
23/9/63-26/9/63	LC

Boilers

New	8683
21/2/40	8826 from 5009
13/10/43	9022 from 5049
11/1/47	9049 from 5014
20/10/51	9046 from 45151
14/10/55	8684 from 45007 (domed)
4/6/60	11982 from 44899 (sloping throatplate)

Tenders

New	9120
6/9/44	9711 (welded)
3/11/45	9122
8/10/51	10679 (part-welded)
14/10/55	10712 (part-welded)
31/1/57	9534 (welded)

Mileage/(weekdays out of service)

Year	Mileage (weekdays out of service)
1935	48,069 (101)
1936	52,302 (48)
1937	43,941 (86)
1938	49,753 (82)
1939	56,415 (44)
1940	45,865 (69)
1941	43,656 (84)
1942	48,324 (69)
1943	53,973 (69)
1944	52,876 (48)
1945	47,120 (71)
1946	44,856 (80)
1947	45,044 (57)
1948	41,682 (71)
1949	35,340 (94)
1950	36,270 (96)
1951	21,755 (151)
1952	44,485 (42)
1953	39,323 (85)
1954	36,820 (88)
1955	38,801 (57)
1956	42,817 (44)
1957	43,705 (63)
1958	40,217 (65)
1959	31,868
1960	35,743
1961	20,167
1962	29,938

Mileage at 12/36: 100,371
Mileage at 31/12/50: 745,486

Sheds

Newton Heath	26/1/35
Bristol	28/9/35
Gloucester	18/1/36
Carlisle N	28/11/36
Inverness	30/5/42
Polmadie	16/2/60

Withdrawn w.e. 22/2/64

5066 at Kings Norton in original Vulcan Foundry condition with tall chimney, prominent topfeed pipes, no cab gutter, plain tender axleboxes and narrow spaced LMS on the tender. It was one of the last to be converted to sloping throatplate, in 1960. Originally allocated to Newton Heath it spent a short time on the Midland Division at Bristol and Gloucester in 1935/36 before going to the Northern Division, where it remained until withdrawn.

45067

Built as 5067 at Vulcan Foundry 5/1/35
Renumbered 45067 w.e. 13/11/48

Improvements and modifications
20/3/38	BTH speed indicator
23/6/38	Removal of vacuum pump
?	Steam sanding
10/1/57	Modification
7/5/59	Fitting BR ATC equipment

Repairs
20/11/35-16/12/35	LS
4/5/36-27/5/36	LO
29/6/36-18/7/36	LO
11/6/37-21/7/37	HG
3/1/38-31/1/38	LO
31/5/38-23/6/38	LS
6/5/39-12/6/39	HG
14/5/40-27/5/40	HS
25/11/41-18/12/41	LS
16/11/42-4/12/42	HG
22/8/44-9/9/44	LS
15/9/45-17/10/45	LS
28/11/46-28/12/46	HG
13/10/48-12/11/48	HS
21/11/49-7/12/49	HI
24/4/51-30/5/51	HG
9/4/53-9/5/53	LI
22/1/55-11/2/55	HI
3/12/56-10/1/57	HG
11/4/59-7/5/59	HI
18/6/59-27/6/59	NCRect (EO)
13/7/60-14/9/60	LC(EO)
2/10/61-28/10/61	LI
16/4/63-11/5/63	HG

Boilers
New	8684
30/6/37	8685 from 5068 (domed)
12/6/39	8982 from 5202
4/12/42	8651 from 5099 (domed)
28/12/46	9037 from 5110
30/5/51	8981 from 45200
10/1/57	8651 from 45106 (domed)

Tenders
New	9121

Mileage/(weekdays out of service)
Year	Mileage (days)
1935	67,134 (95)
1936	57,070 (117)
1937	48,750 (77)
1938	68,194 (80)
1939	50,999 (99)
1940	45,049 (67)
1941	29,103 (74)
1942	28,398 (50)
1943	36,903 (18)
1944	28,213 (48)
1945	31,516 (54)
1946	29,854 (63)
1947	27,888 (63)
1948	23,210 (83)
1949	35,237 (53)
1950	39,844 (39)
1951	39,494 (52)
1952	29,704 (33)
1953	38,375 (47)
1954	42,402 (34)
1955	36,664 (62)
1956	31,066 (61)
1957	40,309 (138)
1958	38,206 (41)
1959	37,787
1960	34,132

Mileage at 12/36: 124,204
Mileage at 31/12/50: 647,362

Sheds
Newton Heath	26/1/35
Trafford Park	10/2/35
Tebay	14/11/36
Patricroft	2/7/37
Kentish Town	7/8/37
Toton	31/5/41
Staveley	7/6/41
Bescot	18/7/42 (loan)
Bescot	15/8/42
Aston	29/8/42
Crewe South	3/10/42
Crewe North	26/6/43
Crewe South	7/8/43
Walsall	3/8/57 (loan)
Crewe South	17/8/57
Crewe North	20/6/64
Bescot	19/9/64
Saltley	26/3/66
Heaton Mersey	28/5/66

Stored
10/1/66-26/4/66

Withdrawn w.e. 21/10/67

Below. **45067 from Crewe South heads a fitted freight at Carpenters Park on 20 July 1963. It had three periods running with a domed boiler, the last from January 1957 to May 1963 but was finally domeless to withdrawal. The AWS was fitted in May 1959 and it had its original tender throughout. Photograph www.rail-online.co.uk**

45068

Built as 5068 at Vulcan Foundry 12/1/35
Renumbered 45068 w.e. 22/1/49

Improvements and modifications
28/4/38	BTH speed indicator
25/5/39	Removal of vacuum pump
?	Steam sanding
7/11/57	Modification
2/12/61	Fitting BR ATC equipment

Repairs
18/1/35-5/2/35	Cost borne by contractors
12/11/35-20/12/35	LS
23/6/36-10/8/36	HO
23/4/37-15/5/37	HG
18/3/38-28/4/38	LS
1/6/38-2/8/38	LO
5/1/39-25/5/39	LS
5/2/40-28/3/40	HG
3/7/41-1/8/41	LS
19/10/41-27/11/41	LO
20/4/42-13/5/42	LO
19/10/42-13/11/42	LS
28/10/43-23/11/43	HG
25/9/44-12/10/44	LS
10/9/45-12/10/45	HS
6/11/46-14/12/46	LS
8/7/47-28/8/47	HG
14/12/48-18/1/49	LI
23/6/50-26/7/50	HI
21/11/51-1/1/52	HG
10/8/53-9/9/53	LI
21/4/54-8/5/54	LC(EO)
30/11/55-10/1/56	HI
1/10/57-7/11/57	HG
2/5/60-7/6/60	LI
21/12/60-17/1/61	LC(EO)

Boilers
New	8685
3/5/37	8823 from 5006 (domed)
28/3/40	9040 from 5110
23/11/43	8637 from 5019 (domed)
28/8/47	9002 from 5207
1/1/52	8679 from 45069 (domed)
7/11/57	8827 from 45064

Tenders
New	9122
12/10/45	9604 (welded)

Mileage/(weekdays out of service)
1935	61,790 (117)
1936	56,010 (120)
1937	47,503 (68)
1938	44,546 (139)
1939	38,728 (148)
1940	47,443 (65)
1941	33,599 (105)
1942	56,756 (59)
1943	47,608 (40)
1944	49,352 (55)
1945	45,460 (67)
1946	39,773 (109)
1947	40,257 (91)
1948	44,426 (63)
1949	49,574 (52)
1950	40,275 (81)
1951	41,327 (66)
1952	52,256 (42)
1953	45,137 (71)
1954	50,192 (48)
1955	39,237 (76)
1956	34,541 (40)
1957	29,493 (102)
1958	42,497 (43)
1959	34,939
1960	26,310

Mileage at 12/36: 117,800
Mileage at 31/12/50: 743,100

Sheds
Newton Heath	26/1/35
Trafford Park	10/2/35
Springs Branch	29/11/36
Kentish Town	7/8/37
Leeds	17/11/45
Bank Hall	22/1/49 (loan)
Bank Hall	5/3/49
Newton Heath	27/1/51 (loan)
Bank Hall	17/2/51
Accrington	5/11/55
Rose Grove	4/3/61
Aintree	25/1/64
Warrington	3/7/65

Withdrawn w.e. 1/1/66

Below. 5068 with tall chimney and prominent top feed pipes. There are no domed covers on the firebox dating the picture before April 1937 when it went into works and emerged with a domed boiler.

45069

Built as 5069 at Vulcan Foundry 19/1/35
Renumbered 45069 w.e. 10/4/48

Improvements and modifications
23/1/39	Removal of vacuum pump
23/1/39	Steam sanding
10/2/55	Modernisation
7/11/59	Fitting BR ATC equipment

Repairs
14/1/36-11/2/36	LS
24/6/36-17/7/36	LO
12/10/36	TRO
27/5/37-11/6/37	HS
21/12/38-23/1/39	HG
15/5/40-29/5/40	LS
26/5/42-13/6/42	LS
13/3/44-27/3/44	HG
12/2/46-2/3/46	HS
26/2/48-7/4/48	HG
23/8/49-20/9/49	HI
25/6/51-8/8/51	HG
19/2/52-12/3/52	LC
2/2/53-27/2/53	HI
2/2/54-6/3/54	LC(EO)
14/1/55-10/2/55	HG
14/2/56-16/3/56	LC
23/5/57-14/6/57	LI
29/4/58-29/5/58	HC
27/9/59-7/11/59	LI
2/8/60-1/11/60	HG
10/1/64-7/2/64	LI

Boilers
New	8686
4/1/39	9056 from 5071
27/3/44	8911 from 5048
7/4/48	8679 from 5048 (domed)
8/8/51	8908 from 45105
10/2/55	8823 from 45041 (domed)
1/11/60	9047 from 45050

Tenders
New	9123
1/11/60	9348
22/2/64	9779 (welded)

Mileage
1935	61,655 (106)
1936	51,462 (130)
1937	43,255 (33)
1938	39,615 (62)
1939	49,188 (63)
1940	36,832 (66)
1941	34,324 (39)
1942	31,807 (46)
1943	37,389 (33)
1944	34,178 (39)
1945	29,343 (45)
1946	34,755 (45)
1947	35,035 (52)
1948	31,906 (76)
1949	27,432 (62)
1950	27,447 (22)
1951	29,880 (64)
1952	38,525 (52)
1953	41,947 (53)
1954	37,890 (53)
1955	43,833 (50)
1956	41,744 (58)
1957	44,312 (54)
1958	42,131 (55)
1959	44,216
1960	36,646

Mileage at 12/36: 113,114
Mileage at 31/12/50: 605,620

Sheds
Newton Heath	26/1/35
Trafford Park	10/2/35
Willesden	19/12/36 (loan)
Preston	2/1/37
Springs Branch	19/6/37
Warrington	13/11/37
Chester	2/7/38
Mold Jcn	24/9/38
Chester	4/2/39
Edge Hill	16/9/39
Chester	30/12/39
Edge Hill	14/12/40
Crewe South	5/9/42
Crewe North	26/6/43
Crewe South	4/3/44
Walsall	5/6/48
Bushbury	2/10/48
Monument Lane	28/5/49
Bescot	1/10/49
Walsall	10/6/50
Bescot	30/9/50
Rugby	9/6/51
Edge Hill	15/3/52 (loan)
Northampton	24/5/52 (loan)
Rugby	5/7/52
Edge Hill	19/9/53
Crewe South	16/9/61
Edge Hill	9/12/61
Holyhead	23/6/62
Speke Jcn	15/9/62
Trafford Park	16/2/63
Springs Branch	22/6/63
Edge Hill	6/7/63

Withdrawn w.e. 10/6/67

Edge Hill allocated 45069 on 20 March 1955 at Willesden, newly fitted with a domed boiler the previous month during a HG repair. It still has its original riveted tender which it kept until 1960.

45069 near Wigan in the late 1950s. It has a domed boiler which it carried between February 1955 and November 1960 and its original riveted tender. The short parcels train includes an SR van followed by LNER, a BR Mark 1 and LMS full brakes. 45069 was at Edge Hill from September 1953 until it moved to Crewe South in 1961.

45070

Built as 5070 at Crewe Works 23/5/35
Renumbered 45070 w.e. 5/3/49

Improvements and modifications
28/11/38	Steam sanding
28/11/38	Removal of vacuum pump
20/1/59	Fitting BR ATC equipment

Repairs
15/6/36-29/6/36	LS
23/4/37-10/5/37	HS
19/10/38-28/11/38	HG
12/4/40-25/4/40	HS
10/4/42-8/5/42	LS
17/11/43-11/12/43	HG
3/9/45-9/10/45	HS
25/9/47-3/11/47	HS
16/2/49-2/3/49	HG
8/6/50-27/6/50	LI
12/10/51-1/11/51	LI
6/2/52-29/2/52	LC
27/6/53-10/8/53	HG
24/4/54-15/5/54	HC(EO)
18/1/55-7/2/55	LI
25/11/56-24/12/56	HI
14/8/58-18/9/58	HG
12/1/59-20/1/59	NC(EO)
15/10/59-27/11/59	HI
24/5/61-21/6/61	HI
30/7/63-30/8/63	HG
23/9/65-23/10/65	LI

Boilers
New	9055
28/11/38	9059 from 5074
11/12/43	8832 from 5009
2/3/49	8925 from 5031
10/8/53	8825 from 45102
18/9/58	8644 from 45102 (domed)
?	8660 (domed)

Tenders
New	9166
5/11/43	9538 (welded)
9/10/45	9711 (welded)
4/7/63	10436 (welded)
?	9480 (welded)

Mileage/(weekdays out of service)
1935	38,365 (24)
1936	56,721 (29)
1937	48,519 (58)
1938	38,569 (82)
1939	48,506 (61)
1940	30,917 (74)
1941	34,868 (51)
1942	38,439 (35)
1943	28,546 (46)
1944	27,854 (24)
1945	29,260 (63)
1946	32,163 (56)
1947	21,343 (68)
1948	37,140 (47)
1949	43,039 (83)
1950	43,256 (53)
1951	38,366 (49)
1952	38,636 (47)
1953	41,299 (66)
1954	41,766 (51)
1955	40,769 (50)
1956	30,531 (47)
1957	42,482 (24)
1958	32,012 (79)
1959	35,711
1960	45,120

Mileage at 12/36: 95,086
Mileage at 31/12/50: 597,505

Sheds
Crewe	25/5/35
Llandudno Jct	17/10/36
Bangor	28/5/38
Rugby	8/4/39
Edge Hill	23/8/41
Patricroft	6/9/41
Crewe South	1/1/44
Aston	8/4/44
Bescot	6/5/44
Willesden	28/10/44
Springs Branch	24/4/48
Chester	8/5/48
Crewe North	21/8/48
Holyhead	28/5/49
Chester	10/2/51
Edge Hill	24/3/51
Aston	8/12/51
Bletchley	3/5/52
Carlisle Upperby	4/7/53
Patricroft	25/6/55
Longsight	9/3/57 (loan)
Patricroft	23/3/57
Carlisle Upperby	21/9/57
Crewe North	30/1/60
Holyhead	11/6/60
Mold Jct	17/9/60
Springs Branch	15/9/62
Warrington	26/6/65

Stored
12/12/66-1/4/67

Withdrawn w.e. 20/5/67

45070 in Crewe Works with newly built BR 2-6-2T 84009. The Black 5 had been repainted in BR lined mixed traffic livery during a Heavy General overhaul completed in August 1953. 45053 was allocated to Bletchley as shown by its 1E shedplate when it entered the works but was officially transferred to Carlisle Upperby while it was in the shops. It has a welded tender and its last domeless boiler, one of the original 14-element boiler type as indicated by the cover plates on the firebox shoulders. Photograph www.rail-online.co.uk

45070 had recently been transferred to Crewe North from Upperby when it was pictured at Stockport on 24 March 1960. With its first domed boiler fitted in September 1958, it had welded tenders since 1943 and AWS in January 1959. Photograph D. Forsyth, Paul Chancellor Collection.

45071

Built as 5071 at Crewe Works 23/5/35
Renumbered 45071 w.e. 7/8/48

Improvements and modifications
?	Steam sanding
?	Removal of vacuum pump
30/4/57	Modification
5/3/59	Fitting BR ATC equipment

Repairs
2/10/36-22/10/36	LS
19/4/37-3/5/37	LO
25/11/37-15/12/37	LS
5/7/38-9/7/38	LO
9/8/38-3/11/38	HG
30/1/40-28/2/40	LS
16/10/41-19/10/41	LO
19/1/42-7/2/42	LS
9/5/42-25/5/42	LO
29/6/43-21/7/43	HG
1/9/44-23/9/44	HS
22/6/45-26/7/45	HO
2/1/47-1/2/47	LS
19/6/48-5/8/48	HG
1/6/50-19/6/50	LI
11/12/51-22/1/52	LI
4/2/52-8/2/52	NC(Rect)
3/2/53-28/2/53	HG
1/5/54-15/5/54	HC(EO)
24/5/55-15/6/55	HI
27/2/56-20/3/56	LC
8/3/57-30/4/57	HG
19/9/58-15/10/58	LC(EO)
24/2/59-5/3/59	NC(EO)
8/12/59-22/1/60	HI
15/11/62-22/12/62	HG

Boilers
New	9056
3/11/38	8996 from 5216
21/7/43	8926 from 5089
5/8/48	8986 from 5034
28/2/53	8941 from 45222
30/4/57	8918 from 45017
?	8906

Tenders
New	9167
11/6/65	9502 (welded)

Mileage/(weekdays out of service)
1935	39,178 (12)
1936	44,815 (70)
1937	40,745 (87)
1938	35,271 (153)
1939	52,711 (89)
1940	31,530 (58)
1941	31,182 (74)
1942	37,406 (57)
1943	25,271 (55)
1944	32,112 (50)
1945	20,328 (90)
1946	40,326 (32)
1947	29,544 (75)
1948	30,614 (73)
1949	43,258 (35)
1950	34,948 (54)
1951	34,751 (53)
1952	34,217 (76)
1953	41,532 (61)
1954	43,910 (47)
1955	38,558 (55)
1956	45,742 (47)
1957	35,027 (102)
1958	33,546 (90)
1959	31,208
1960	38,166

Mileage at 12/36: 83,993
Mileage at 31/12/50: 569,239

Sheds
Crewe	25/5/35
Northampton	7/3/36
Rugby	24/7/37
Millhouses	2/4/38 (Loan)
Millhouses	9/4/38
Trafford Park	7/1/39
Toton	9/3/40
Rugby	18/7/42 (loan)
Rugby	15/8/42
Crewe South	3/10/42
Willesden	13/11/43
Stoke	3/5/47
Willesden	10/5/47
Bletchley	3/5/51
Crewe North	13/8/53
Carlisle Upperby	10/9/55
Monument Lane	22/6/57
Bushbury	17/9/60
Monument Lane	17/6/61
Speke Jcn	28/10/61

Stored
27/2/38-26/3/38

Withdrawn w.e. 22/7/67

Below. **45071 at Beeston Castle in April 1954 with a wonderful mixed train with two horseboxes and two cattle vans at the front. It had been transferred to Crewe North the previous year and was one of the class to always carry a domeless boiler. Photograph G.W. Sharpe.**

45072

Built as 5072 at Crewe Works 27/5/35
Renumbered 45072 w.e. 2/7/49

Improvements and modifications
19/8/38	Removal of vacuum pump
6/2/43	Steam sanding
5/9/57	Modernisation
13/11/59	Fitting BR ATC equipment

Repairs
11/5/36-25/5/36	LS
19/4/37-4/5/37	HS
15/7/38-19/8/38	HG
1/10/39-23/10/39	TRO
13/2/40-17/2/40	HS
11/12/41-10/1/42	LS
22/12/42-6/2/43	HG
10/2/44-16/3/44	LO
23/5/44-3/6/44	LO
13/2/45-10/3/45	HS
2/12/47-31/12/47	HS
2/6/49-28/6/49	HG
29/11/49-16/12/49	LC
15/2/50-28/2/50	NC
8/5/50-26/5/50	LC
19/5/51-7/6/51	LI
28/3/53-5/5/53	HG
14/12/53-7/1/54	LC(EO)
5/3/55-1/4/55	HI
9/9/55-5/10/55	LC(EO)
29/10/55-19/11/55	LC(EO)
30/7/57-5/9/57	HG
14/10/59-13/11/59	HI
28/3/61-3/5/61	LI
7/11/63-16/12/63	HG

Boilers
New	9058
19/8/38	8908 from 5128
6/2/43	8830 from 5011
10/3/45	8821 from 5224 (domed)
28/6/49	8678 from 5145 (domed)
5/5/53	8986 from 45071
5/9/57	9031 from 45104
?	8644 (domed)

Tenders
New	9168

Mileage/(weekdays out of service)
1935	42,466 (32)
1936	51,571 (31)
1937	43,348 (45)
1938	44,620 (54)
1939	35,117 (72)
1940	35,610 (32)
1941	28,442 (73)
1942	30,771 (55)
1943	33,570 (50)
1944	29,030 (90)
1945	35,586 (41)
1946	28,459 (58)
1947	29,424 (81)
1948	39,138 (33)
1949	31,326 (72)
1950	32,755 (62)
1951	46,690 (49)
1952	44,947 (45)
1953	44,241 (79)
1954	49,465 (38)
1955	41,624 (85)
1956	45,540 (46)
1957	37,965 (65)
1958	44,267 (40)
1959	38,676
1960	45,639

Mileage at 12/36: 94,037
Mileage at 31/12/50: 571,223

Sheds
Crewe	1/6/35
Edge Hill	7/3/36
Crewe South	27/6/42
Crewe North	26/6/43
Crewe South	4/3/44
Warrington	1/5/48
Carnforth	14/10/50
Rugby	19/4/58 (loan)
Carnforth	5/7/58
Edge Hill	14/5/60
Holyhead	11/6/60
Mold Jnc	17/9/60
Carlisle Upperby	10/11/62
Carnforth	22/6/63

Withdrawn w.e. 9/9/67

Below. 45072 in the 1960s at Carlisle Kingmoor with a domed boiler which was almost certainly fitted in December 1963. It was allocated to Carnforth from July 1963 until withdrawn in September 1967. It received AWS in November 1959 and kept its original riveted tender throughout. Photograph www.rail-online.co.uk

45073

Built as 5073 at Crewe Works 5/6/35
Renumbered 45073 w.e. 21/8/48

Improvements and modifications
19/4/38	Removal of vacuum pump
12/10/44	Steam sanding
21/2/59	Fitting BR ATC equipment

Repairs
29/9/36-16/10/36	LS
4/6/37-17/6/37	LO
15/2/38-19/4/38	HG
27/10/39-18/11/39	LS
5/3/41-12/3/41	LO
28/7/41-23/8/41	HG
17/9/42-30/9/42	LO
21/7/43-13/8/43	LS
23/9/44-12/10/44	HG
24/9/46-12/10/46	LS
20/7/48-21/8/48	HG
5/1/50-6/2/50	HI
5/11/51-6/12/51	HI
2/10/53-28/10/53	HG
4/9/55-1/10/55	LI
17/11/56-14/12/56	LI
5/8/57-14/9/57	LO(EO)
26/7/58-23/9/58	HG
2/2/59-21/2/59	NC(EO)
8/8/59-20/11/59	LC(EO)
5/7/60-15/9/60	HI

Boilers
New	9057
4/4/38	8966 from 5186
23/8/41	8971 from 5186
12/10/44	9016 from 5154
21/8/48	8834 from 5102
28/10/53	8926 from 45056
23/9/58	8910 from 45222

Tenders
New	9000 (prototype)
12/10/44	9237 (welded)

Mileage/(weekdays out of service)
1935	39,123 (30)
1936	46,762 (44)
1937	39,113 (57)
1938	39,182 (74)
1939	35,892 (58)
1940	32,525 (36)
1941	28,427 (53)
1942	29,882 (34)
1943	35,023 (58)
1944	29,747 (49)
1945	35,189 (36)
1946	33,566 (46)
1947	33,764 (57)
1948	32,253 (67)
1949	35,297 (49)
1950	31,693 (60)
1951	30,742 (61)
1952	41,782 (35)
1953	36,182 (60)
1954	44,278 (23)
1955	36,267 (55)
1956	33,184 (51)
1957	42,585 (61)
1958	41,349 (78)
1959	33,972
1960	35,010

Mileage at 12/36: 85,885
Mileage at 31/12/50: 557,438

Sheds
Crewe	8/6/35
Aston	7/3/36
Bescot	19/4/41
Crewe North	10/5/41
Carlisle Upperby	9/2/57
Aston	3/8/57
Carlisle Upperby	17/8/57
Crewe North	19/10/57
Lancaster	10/6/61
Speke Jcn	9/9/61
Springs Branch	21/7/62
Newton Heath	27/7/63
Trafford Park	28/11/64
Stockport	28/10/67
Bolton	13/4/68
Lostock Hall	6/7/68

Stored
31/10/66-28/10/67

Withdrawn w.e. 3/8/68

One of the class to survive until the end of BR steam in August 1968 45073 waits at Preston in 1963/4. It was shedded at 9D Newton Heath from July 1963 to November 1964 and was fitted with AWS in February 1959. Always domeless, 45073 was built with one of the three prototype Stanier tenders which it kept until 1944 when it received the welded tender which lasted until withdrawn. Photograph www.rail-online.co.uk

45074

Built as 5074 at Crewe Works 12/6/35
Renumbered 45074 w.e. 3/7/48

Improvements and modifications
16/9/38	Removal of vacuum pump
4/8/45	Steam sanding
17/9/56	Modernisation
25/4/59	Fitting BR ATC equipment

Repairs
28/10/36-16/11/36	HS
11/1/38-26/1/38	HS
12/8/38-16/9/38	HG
16/4/40-30/4/40	LS
25/7/41-14/8/41	HG
3/4/42-18/4/42	LO
19/4/43-5/5/43	HS
5/9/44-20/9/44	LS
25/6/45-4/8/45	HG
16/10/47-29/11/47	LS
2/6/48-3/7/48	LO
13/9/48-19/10/48	HO
10/8/49-12/10/49	HG
31/7/51-25/8/51	LI
2/10/52-5/11/52	HG
27/8/54-22/9/54	LI
7/8/56-17/9/56	HG
26/9/56-10/10/56	NC(Rect)(EO)
31/3/59-25/4/59	HG
29/5/59-1/6/59	NC(Rect)(EO)
12/5/59-21/5/59	LC(EO)
25/2/60-14/4/60	LC(EO)
19/9/62-13/10/62	LI

Boilers
New	9059
16/9/38	8980 from 5200
14/8/41	8641 from 5136 (domed)
4/8/45	9045 from 5175
12/10/49	8821 from 5072 (domed)
5/11/52	9033 from 45189
17/9/56	8972 from 45220
25/4/59	8921 from 45051

Tenders
New	9001 (prototype 4000 gallon)
4/8/45	9208 (riveted)

Mileage/(weekdays out of service)
1935	32,491 (12)
1936	47,225 (45)
1937	41,196 (41)
1938	39,199 (79)
1939	47,293 (34)
1940	32,325 (31)
1941	36,911 (48)
1942	31,225 (54)
1943	37,914 (45)
1944	31,363 (47)
1945	32,605 (70)
1946	32,942 (34)
1947	29,202 (83)
1948	28,627 (94)
1949	29,814 (89)
1950	38,799 (35)
1951	35,669 (40)
1952	35,631 (57)
1953	44,398 (34)
1954	36,158 (53)
1955	35,282 (67)
1956	27,919 (80)
1957	42,794 (23)
1958	40,006 (52)
1959	35,868
1960	33,627

Mileage at 12/36: 79,716
Mileage at 31/12/50: 569,121

Sheds
Crewe	15/6/35
Aston	7/3/36
Longsight	4/6/38
Chester	2/7/38
Longsight	9/7/38
Stockport	30/9/39
Stoke	6/1/40
Crewe North	29/3/41
Crewe South	23/10/41
Crewe North	26/6/43
Crewe South	4/3/44
Bristol	1/7/50 (loan)
Sheffield	5/8/50 (loan)
Crewe South	11/11/50
Stoke	15/12/62

Withdrawn w.e. 18/9/65

Below. Allocated to Crewe South until the end of 1962, 45074 slogs up the hills with a fitted freight in the early 1960s. It acquired a welded tender in 1945, AWS in May 1959 and had a domeless boiler from 1952 onwards. It exchanged its original prototype Stanier tender for this riveted example in 1945. Photograph R.K. Blencowe.

Endpiece

Holyhead's 45043 at Bangor in June 1966 shows in detail two fittings not usually this clear in photographs. Firstly the draught shield on the cab window and secondly the clip to hold a train staff above the fourth digit of the cab number. Photograph N. Kneale, www.transporttreasury.co.uk